Black Liquor Gasification

Black Liquor Gasification

Pratima Bajpai

AMSTERDAM · BOSTON · HEIDELBERG · LONDON · NEW YORK
OXFORD · PARIS · SAN DIEGO · SAN FRANCISCO · SYDNEY · TOKYO

Elsevier
32 Jamestown Road, London NW1 7BY
225 Wyman Street, Waltham, MA 02451, USA

First published 2014

British Library Cataloguing in Publication Data
A catalogue record for this book is available from the British Library

Library of Congress Cataloging-in-Publication Data
A catalog record for this book is available from the Library of Congress

ISBN: 978-0-08-100009-0

For information on all Elsevier publications
visit our website at **store.elsevier.com**

This book has been manufactured using Print On Demand technology. Each copy is produced to order and is limited to black ink. The online version of this book will show color figures where appropriate.

CONTENTS

PREFACE

Black liquor gasification (BLG) is currently being developed as an alternative technology for energy and chemical recovery. Black liquor is particularly rich in energy content and is already used to make energy for the mill, but energy generation from by-products such as black liquor has traditionally been a low economic priority relative to process efficiencies in making pulp. That priority is changing as energy value grows, yet the compelling reasons to sustain process efficiency remain. BLG is a possible way to improve the efficiency of converting black liquor biomass into energy and it operates largely in sync with other methods to improve pulp-making efficiency. BLG offers a way to generate electricity and to reclaim pulping chemicals from black liquor. This is accomplished by converting the fixed carbon to a combustible gas mixture using oxygen-containing gases such as oxygen, carbon dioxide, and water vapor. The combustible gas is then burned to generate electrical power. BLG would replace the Tomlinson recovery boiler for the recovery of spent chemicals and energy. Gasification may become part of integrated gasification and combined cycle (IGCC) operation, or lead to pulp mills becoming biorefineries. This e-book concerns the gasification of black liquor. It looks at the technical history of BLG and the obstacles to its wide adoption by pulp mills. Finally, it reviews the economic and market concerns likely to drive adoption in different markets.

ABBREVIATIONS

BLG	black liquor gasification
BLGCC	black liquor gasification combined cycle
BLGMF	black liquor gasification motor fuel
DME	dimethyl ether
DP1	Demonstration Plant 1
DARS	Direct Alkali Recovery System
FT	Fisher−Tropsch
HRSG	heat recovery steam generator
HTBLG	high temperature black liquor gasification
HTBLGCC	high temperature black liquor gasification combined cycle
LP	low pressure
LTBLG	low temperature black liquor gasification
MP	medium pressure
MTCI	Manufacturing Technology Conversion International Company
RB	recovery boiler
TRI	ThermoChem Recovery International Company
WTW	well-to-wheel

General Background

In the pulp and paper industry, large quantities of forest biomass are being used. The by-products or residues which result include black liquor, bark, and forest logging residues. These can be used for energy purpose to produce electricity, heat, and biofuels. The kraft pulping process accounts for almost 60% of all pulp production (Joelsson and Gustavsson, 2008; Holmberg and Gustavsson, 2007). Wood chips are cooked at high temperature and pressure using white liquor to dissolve lignin. The spent cooking liquor, called black liquor, contains inorganic cooking chemicals and combustible material. An integral part of kraft process is to recover cooking chemicals and energy from black liquor in the recovery boilers known as the recovery cycle. Chemical pulp mills around the world have been relying on Tomlinson recovery furnaces to process spent pulping liquors and produce by-product steam for process use and electrical cogeneration. In general, the technology has served the industry well. Although evolutionary developments have kept Tomlinson-anchored chemical recovery "islands" or loops performing more or less satisfactorily since the 1930s, problems inherent with an outdated, aging technology continue to nag mills and drag down overall pulp mill process efficiency. Table 1.1 shows the shortcomings of the recovery boiler.

Among recovery boiler problems that seem to never go away is related to production bottlenecks in the pulp mill. Many mills today operate at or near existing chemical recovery capacity, and over the years a growing number have become recovery boiler limited to the

Table 1.1 Shortcomings of the Recovery Boiler
Relatively low energy efficiency
Relatively poor environmental performance
Challenging boiler control
Difficult to control mill sulfur balance
Risk for explosion

tune of 100–500 tpd of black liquor solids (BLS). Because the addition of recovery furnace capacity at these smaller incremental levels is not practical, such mills are faced with continuing pressure to operate below maximum efficiency, curtailing production, or investing many millions of dollars to install a new boiler. Many recovery furnaces around the world are nearing or have surpassed their functional lifespans and not only are becoming progressively inefficient in their old age but also represent a potential explosion danger due to increasing metal fatigue, corrosion, stress cracking, etc. Clearly, mills face some critical capital spending decisions in the future. In 1999, Jaako Pöyry conducted a study to assess current and future capacity requirements and number of recovery boilers, which require upgradation or replacement in near future. According to the study, many recovery boilers around the world, especially in United States, are near to complete their functional lifespans. Majority of these recovery boilers were built in the late 1960s through the 1970s (Patrick, 2003).

Black liquor from kraft process represents a potential energy source of 250–500 MW per mill. According to Food and Agriculture Organization in United Nations, the pulp and paper industry around the world currently processes more than 215 million tons of BLS per year, with a total energy content of about 2 EJ (lower heating value (LHV), 12.3 MJ/kg as BLS), which represents about 0.4% of the world marketed energy consumption, i.e., approximately 510 EJ in 2008 (U.S. EIA, 2009). This large amount of energy makes black liquor a very significant biomass fuel (Naqvi et al., 2010). In fact, black liquor has been the main biomass resource as bioenergy in some countries with large pulp and paper industry. Today, this large amount of energy is used to meet process energy demands of the pulp mills and still a part of total electricity demand is fulfilled by importing electricity. If black liquor is gasified instead of combustion in the recovery boiler, a pulp mill can shift from electricity importer to electricity exporter using

Table 1.2 Advantages of BLG Technology Compared to the Existing Recovery System
An enlarged capacity for electric energy recovery
An enlarged capacity for chemicals recovery
A natural splitting of sulfur and alkali streams leading to the possibility of application in advanced pulping processes, increasing the yield of pulp
The exclusion of water-smelt explosions due to absence of water or steam tubes inside the reactor
Cleaner production from an environmental point of view (emission reduction of nitrogen oxides (NO_x) and sulfur oxides (SO_x))
A cut in servicing expenditures

BLGCC (black liquor gasification combined cycle) technology. The organic constituents of the black liquor also offer synthesis gas for biofuel production such as hydrogen, methanol, dimethyl ether (DME), and methane.

The industry has been exploring alternative chemical recovery technologies to replace or augment the conventional Tomlinson boiler system since the 1970s. Among these, black liquor gasification (BLG) has shown some interesting promise. This technology has gone through a stepwise development since its early predecessor was developed in the 1960s. It is one of the high-prioritized R&D areas and is considered as an alternative technology to replace conventional recovery cycle with the recovery boiler. This topic has been popular in several conferences on biorefining, engineering, pulping, and environmental matters (Andersson and Harvey, 2004; Ådahl et al., 2004; Bajpai, 2008, 2012; Berglin et al., 1999, 2002, 2003; Dahlquist, 2003; Dahlquist and Jacobs, 1994; Dahlquist and Jones, 2005; Dahlquist et al., 2009; Ekbom et al., 2003; Harvey and Facchini, 2004; Möllersten et al., 2003a,b, 2004; Maunsbach et al., 2001; Yan and Eidensten, 2000; Yan et al., 1995; Jonsson and Yan, 2005; Eriksson and Harvey, 2004; Näsholm and Westermark, 1997; Waldner and Vogel, 2005; Sricharoenchaikul, 2009). Table 1.2 shows advantages of BLG technology.

BLG is a process in which a clean synthesis gas (syngas) is produced from black liquor by converting its biomass content into gaseous energy carrier. The syngas subsequently can be used in boilers or in combined cycle processes (utilizing gas turbines) to generate on-site electricity and/or process steam. The gasification produces a fuel gas that needs to be cleaned to remove undesired impurities for the power system and to recover pulping chemicals. The potential advantages of BLG are the greater end use flexibility offered by a gaseous fuel, reduced air pollutant content, and higher electricity-to-heat ratios in combined cycle systems than standard recovery boiler steam turbine systems. Potential disadvantages of gasification combined cycle systems include the energy investments required for achieving sufficient BLS concentration and higher lime kiln and causticizer loads (and associated fuel inputs) compared to Tomlinson recovery boiler systems. Additionally, since combined cycle systems generate electrical power more efficiently than steam turbine-based systems, meeting the facility's steam demand may require the use of more fuel. However, this additional fuel use also results in more available electricity for facility use or export to the grid. BLG technologies and applications are in continuous states of research and development. The potential benefits and costs of BLG—both environmental and economic—are likely to depend highly on the characteristics of individual installations and will be better understood as the technologies and applications are demonstrated and evaluated over time.

The developments in gasification technology have been conducted over years for efficient recovery of bio-based residues in the chemical pulp mills (Dahlquist et al., 2009; Ekbom et al., 2003; Harvey and Facchini, 2004; Möllersten et al., 2003a,b; Yan et al., 2007; Ådahl et al., 2004). Several studies have been conducted to analyze the technical, economic, and climate change mitigation performance of gasification process and a number of pilot plants have been successfully operated. There has been more focus on possible integration of gasification process for increased energy self-sufficiency. This serves as a base for a modern biorefinery concept at the pulp mills coproducing pulp and valuable energy products.

BLG is a bioenergy technology that has been vigorously researched for more than two decades. The promise of commercial BLG technology appears to be high and increasing. Commercial BLG can meet another 3–4% of US industrial electricity demand through combined cycle power generation depending on differences in technology adoption rates (Larson et al., 2003). This adds to the nearly 3% already produced today at pulp mills by converting black liquor to process steam in boilers. Together, then, replacing the boiler with a gasifier could account for 6–7% of all US industrial energy demand, and these numbers grow for Nordic countries, where mills are newer and overall demand is lower due to higher efficiencies. Estimates suggest that BLG could supply as much as 7% of Sweden's electricity demand, not just its industrial demand, or startlingly, as much as 30% of the entire nation's need for transportation fuels (www.ec.europa.eu/environment/etap). China and Indonesia also display an active interest in BLG to meet their escalating energy needs (Farmer and Sinquefield, 2009). The production of a larger volume of electric energy from BLG is an attractive goal partly because BLG also includes considerable environmental benefits. Yet this has been almost eclipsed by an even more ambitious vision: the installation of a full-scale biorefinery where pulp mills generate multiple value streams—not only electricity but also motor fuels (MFs) specialty chemical products and rare polymers extracted before or during the pulping process (Farmer and Sinquefield, 2009). Essentially, a mature BLG technology supports many of these expensive biorefinery models, especially those that use BLG syngas as an intermediate product to produce MFs, hydrogen for peroxide bleaching or for fuel cells, and mixed alcohols or special high-yield liquors. These are powerful economic motivations despite technical frustrations. Several commercial-scale (100–400 tpd BLS) gasification units have been started in the United States along with a smaller demonstration unit (20 mtpd of solids) at a Kappa Kraftliner mill in Sweden (Patrick, 2003).

1.1 THE PULP AND PAPER INDUSTRY

The pulp and paper industry is one of the largest industry in the world. It is also an important source of employment in many countries. A sustainably managed pulp and paper industry can bring several benefits to the local economy and people particularly in rural areas. The industry is dominated by North American, Northern European, and East Asian countries like Japan. Latin America and Australasia also have significant pulp and paper industries. Over the next few years, it is expected that both India and China will become the key in the industry's growth. The largest producer countries, United States, China, Japan, and Canada, make up more than half of the world's paper production which is 400 million tons a year. Table 1.3 shows key paper consumption facts (www.panda.org).

Table 1.3 Key Paper Consumption Facts

- 50% of the paper and board produced globally is used for packaging
- The second largest market for paper is printing and writing
- 400 million tons per year: Global paper consumption as of 2010. Half of this is consumed by Europeans and North Americans and is thrown away after a short time
- 500 million tons: Forecasted increase in paper consumption by 2020

Table 1.4 Annual Production of Selected Fuels

Production (million tons per year)	
Black liquor solids	200
Paper and board	180
Crude oil	4000
Hard coal	500
Source: Based on Farmer and Sinquefield (2009).	

Pulp and paper are primarily made out of wood fibers originating from natural forests or pulpwood plantations (Biermann, 1996a). Recycled fiber and other fiber sources such as agricultural residue are also utilized and recycled fiber is becoming more commonly used in pulp and paper making. Paper is a versatile product with many end uses varying from household papers, graphic and office papers to medical papers. There are two significant pulping technologies available that differ greatly in terms of process, i.e., mechanical and chemical pulping. Approximately 30% of the total pulp production in European Union is from mechanical pulping while the rest is produced by means of chemical pulping (Swedish Forest Agency, 2008). North America has major pulp and paper industry, about 21% of the total pulp produced is from mechanical pulping and rest is produced chemically.

A pulp mill that produces bleached kraft pulp generates 1.7–1.8 tons of black liquor (measured as dry content) per ton of pulp. Black liquor thus represents a potential energy source of 250–500 MW per mill. As modern kraft pulp mills have a surplus of energy, they could become key suppliers of renewable fuels in the future energy system. Today, black liquor is the most important source of energy from biomass. It is thus of great interest to convert the primary energy in the black liquor to an energy carrier of high value. Table 1.4 gives the relative annual production of some major fuels. A key advantage of black liquor compared to biofuels and fossil fuels is that it is already at the mill; the handling infrastructure already exists and there are no collection and transport costs.

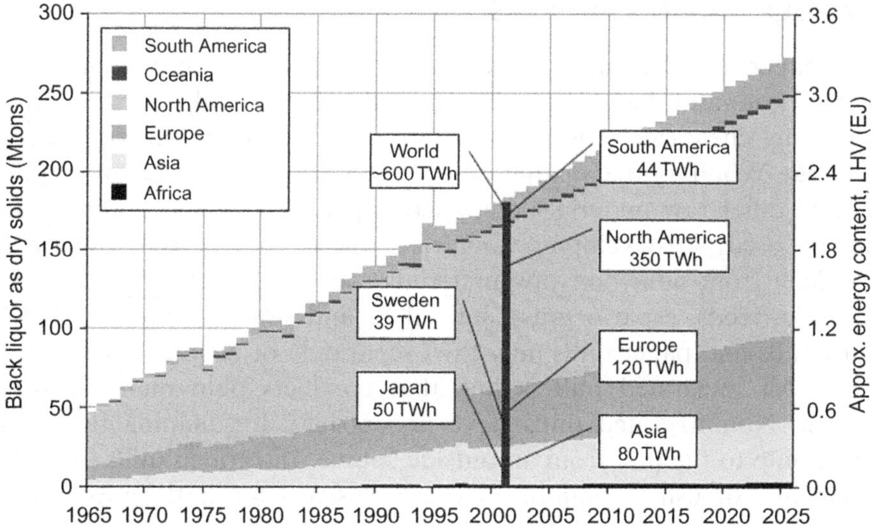

Figure 1.1 Estimated black liquor world production. Reproduced with permission from IEA Bioenergy Report (2009).

Worldwide, the pulp and paper industry currently processes about 170 million tons of black liquor (measured as dry solids) per year, with a total energy content of about 2 EJ, making black liquor a very significant biomass source (IEA Bioenergy Report, 2009) (Figure 1.1). In comparison with other potential biomass sources for chemical production, black liquor has the great advantage that it is already partially processed and exists in a pumpable, liquid form. Using black liquor as a raw material for liquid or gaseous biofuel production in a biorefinery has many advantages: biomass logistics are extremely simplified as the raw material for fuel production is handled within the boundaries of the pulp and paper plant; the process is easily pressurized, which increases fuel production efficiency; due to the processing of wood to pulp, the produced syngas has a low methane content, which optimizes fuel yield; pulp mill economics become less sensitive to pulp prices when diversified with another product; and gasification capital cost is shared between recovery of inorganic chemicals, steam production, and syngas production. Overall, if the global production of black liquor was to be used for transport biofuel production, then this would correspond to about 48 million tons of methanol, compared with current world production from fossil fuels of about 32 million tons, a significant impact.

1.2 PULP MAKING PROCESS

The paper manufacturing process has several stages: raw material preparation and handling, pulp manufacturing, pulp washing and screening, chemical recovery, bleaching, stock preparation, and papermaking. Wood is the main raw material used to manufacture pulp though other raw materials—recycled paper, and agricultural residues can be used. In developing countries, about 60% of cellulose fibers originate from nonwood raw materials such as bagasse, cereal straw, bamboo, reeds, esparto grass, jute, flax, and sisal (Gullichsen, 2000). Pulp mills and paper mills may exist separately or as integrated operations. An integrated mill is one that conducts pulp manufacturing on-site. Nonintegrated mills have no capacity for pulping but must bring pulp to the mill from an outside source. Integrated mills have the advantage of using common auxiliary systems for both pulping and papermaking such as steam and electric generation and wastewater treatment. Transportation cost is also reduced. Nonintegrated mills require less land, energy, and water than integrated mills. Their location can, therefore, be in a more open setting where they are closer to large work force populations and perhaps to their customers. Wood typically enters a pulp and paper mill as logs or chips and is processed in the wood preparation area referred to as the woodyard. In general, woodyard operations are independent of the type of pulping process. If the wood enters the woodyard as logs, a series of operations converts the logs into a form suitable for pulping, usually wood chips. Logs are transported to the slasher, where they are cut into desired lengths, followed by debarking, chipping, chip screening, and conveyance to storage. The chips produced from logs or purchased chips are usually stored on-site in large storage piles. Most of the pulp and paper produced today (90%) originates from wood. The major components of softwoods (e.g., pine, spruce) as well as hardwoods (e.g., birch, aspen, eucalypts) are cellulose (40–50%), hemicellulose (25–30%), and lignin (25–30%). Extractives constitute a minor part.

Pulp for paper production is obtained via mechanical pulping and chemical pulping. These two processes differ greatly in principle:

1. *Mechanical pulping*, in which the fibers are separated mainly through mechanical treatment in refiners. Most of the wood thus becomes pulp, including the lignin.

2. *Chemical pulping*, in which the fibers are separated mainly through chemical treatment in either acidic or caustic solutions. These processes aim to separate the lignin from the cellulose fibers.

In the mechanical pulping, the objective is to maintain the main part of the lignin in order to achieve high yield with acceptable strength properties and brightness. Mechanical pulps have a low resistance to aging which results in a tendency to discolor. Mechanical pulping separates fibers from each other by mechanical energy applied to the wood matrix causing the bonds between the fibers to break gradually and fiber bundles, single fibers and fiber fragments to be released (Smook, 1992; Biermann, 1996b).The main processes are stone groundwood pulping (SGW), pressure groundwood pulping (PGW), thermomechanical pulping (TMP), or chemi-thermomechanical pulping (CTMP). Mechanical pulps are weaker than chemical pulps, but cheaper to produce (about 50% of the costs of chemical pulp) and are generally obtained in the yield range of 85–95%. Currently, mechanical pulps account for 20% of all virgin fiber material. Due to their high lignin content, mechanical pulps quickly become yellow. They are therefore used mostly for products with a short lifespan such as newsprint and magazine paper. Another reason why mechanical pulps are used in these products is because they contain large fractions of relatively short fibers and fiber fragments; therefore, they make dense and opaque sheets that are suitable for printing paper. Approximately one-third of the pulp produced in the European Union is mechanical pulp.

Chemical pulps constitute the other two-thirds of the pulp production. These pulps are characterized by high strength and, if bleached, by high brightness and long-term brightness stability. Typical products made from bleached chemical pulp include fine paper, tissue, and a number of board grades. Unbleached chemical pulp is mostly used to produce corrugated board and sack paper (Smook 1992; Biermann, 1996b). The present chemical pulping processes descend back to the sodium carbonate process initially taken in operation during the 1870s (Kassberg, 1998). At that time, no recovery process was used at all. The residue liquors were simply disposed to nearest watercourse, resulting in environmental problems and raw material costs. One of the first devices constructed to burn the black liquor and recover the cooking chemicals was the flame furnace. It was a batch furnace where the bottom was covered by black liquor. After the liquor was dried by the hot smoke gases from wood firing, it was manually removed out on the factory floor where the final combustion took place. High wood fuel consumption and difficult working conditions were the consequences. Improvement was made when a smelt furnace was built in direct connection to the flame furnace (Råberg, 2007a,b). In the 1880s, sulfur was introduced in the cooking chemicals and the sulfate process was born which gave rise to the characteristic smell of a sulfate pulp mill. The kraft process, uses sodium hydroxide (NaOH) and sodium sulfide (Na_2S) to pulp wood, is the dominant pulping process in the pulp and paper industry. In the kraft process, about half of the wood is dissolved and together with the spent pulping chemicals forms a liquid stream called weak black liquor (Figure 1.2) (Tran and Vakkilainen, 2007). About 130 million tons per year of kraft pulp are produced globally, accounting for two-thirds of the world's virgin pulp production and for over 90% of chemical pulp (Tran and Vakkilainen, 2007). The main advantages of kraft pulping are shown in Table 1.5.

The main disadvantages of kraft cooking are the emission of malodorous gasses, reduced pulp yield as compared with sulfite cooking,

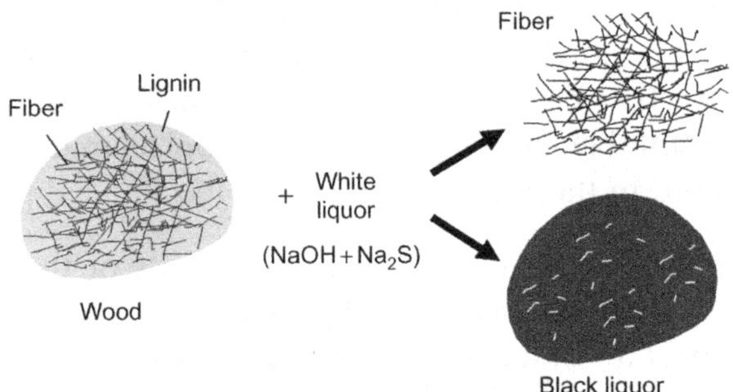

Figure 1.2 *The kraft process.* Reproduced with permission from Tran and Vakkilainen (2007).

Table 1.5 Main Advantages of Kraft Cooking
Lower demand for wood species and quality of wood raw materials, enabling the use of all types of wood and allowing the presence of significant amounts of extractive substances, rotten wood, and bark residues
Well-developed processes of recycling waste liquor, including the regeneration of cooking chemicals and generation of steam
Excellent mechanical properties of pulp
Production of valuable by-products, such as tall oil and turpentine in the cooking of pine wood
Relatively short cooking time
Source: *Based on Gullichsen (2000) and Smook (1992).*

and the dark color of unbleached pulp. In Figure 1.3, a schematic shows a modern kraft mill that produces bleached market pulp. First wood chips are fed into a digester where the cellulose fibers are separated from the wood immersion in hot white liquor. White liquor is a strong aqueous solution consisting of sodium sulfate and sodium sulfide that neutralizes the organic acid and the bounding lignin in wood, and separates the fibers that later become pulp. The pulp is used to produce paper and the remaining by-products (lignin, cooking chemical, etc.) leave the digester as weak "black liquor." Sulfide has two positive effects; it both reduces the reaction rate for carbohydrate dissolution and increases the delignification rate. The drawback is that small amounts of sulfide react to produce organic sulfur compounds such as methyl mercaptan and dimethyl sulfide. The odor threshold for these compounds is very low, and despite the efficiency of modern collection systems for odorous gases there is always a characteristic smell from a kraft pulp mill, which may be almost negligible under continuous trouble-free operation, but becomes evident during upsets or accidental spills. The pulp produced in the digester is washed to recover the cooking liquor and reduce the carryover of dissolved organic material to the oxygen delignification stage. This stage is more selective than cooking, i.e., the yield loss is smaller per unit of lignin removed. After further washing, the pulp goes to the bleach plant. Final bleaching is still more selective than oxygen delignification and is usually

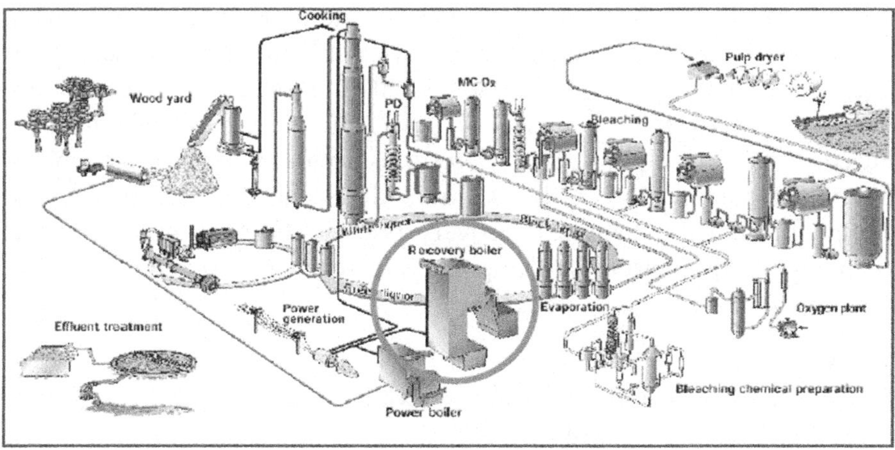

Figure 1.3 Schematic of a modern kraft pulp mill with its process units. In a BLG system only the recovery boiler (marked with red circle) has to be replaced. (For interpretation of the references to color in this figure legend, the reader is referred to the web version of this book.) Reproduced with permission from IEA Bioenergy Report (2009).

done in a sequence of acidic and alkaline stages with washing between the stages. The most common bleaching chemicals used today are chlorine dioxide and hydrogen peroxide. After final bleaching the lignin content is very low, giving the pulp high brightness stability. In a market pulp mill, the bleached pulp is dried with hot air in a pulp dryer before it is baled and shipped to the customers (paper mills). In an integrated mill, the pulp is not dried but pumped to the paper machine, where it can be mixed with other pulps and additives to give the paper its desired properties. Even integrated mills sometimes produce market pulp, because the optimal size of a pulp mill is larger than that of a paper machine.

1.3 RECOVERY CYCLE

Recovery cycle is one of the most important processes in a kraft pulp mill. Half of the wood raw material is utilized as chemical pulp fiber. The other half is utilized as fuel for electricity and heat generation. In fact, a pulp mill has two main lines. Wood is turned into pulp on the fiber line. Energy is produced on the chemical recovery line from the wood material cooked in the liquor; the cooking chemicals are recovered for reuse. In the chemical recovery line, black liquor is evaporated and combusted in a recovery boiler, and the energy content of the dissolved wood material is recovered as steam and electricity. The chemical pulping process generates more energy than it uses. A pulp mill generates energy for its own use and energy to sell (Tran and Vakkilainen, 2007; Vakkilainen, 2000; Bajpai, 2008; Biermann, 1996b; Adams, 1992, 1997; Reeve, 2002). For every ton of pulp produced, the kraft pulping process produces about 10 tons of weak black liquor or about 1.5 tons of black liquor dry solids that need to be processed through the chemical recovery process. Figure 1.4 shows kraft chemical recovery process. The process of regeneration is composed of three operations:

1. Evaporation of black liquor
2. Combustion of condensed liquor, which results in steam and mineral residue in the form of smelt
3. Causticizing of dissolved smelt (green liquor).

The process has three main functions as shown in Table 1.6.

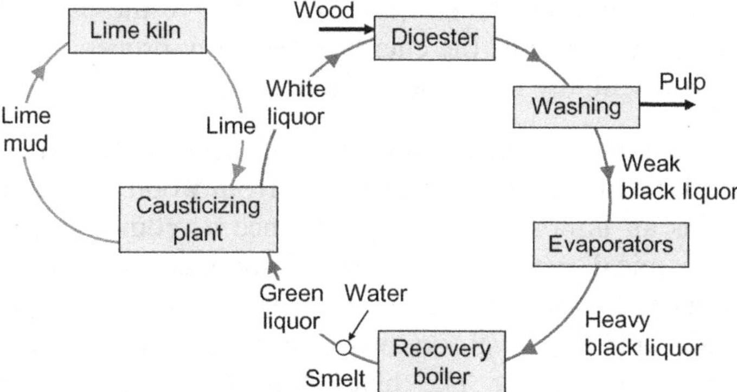

Figure 1.4 Simplified flow diagram of the chemical recovery cycle in the kraft pulping process. Reproduced with permission from Tran and Vakkilainen (2007).

Table 1.6 Main Functions of Kraft Chemical Recovery Process
Minimizing the environmental impact of waste material (black liquor) from the pulping process
Recycling pulping chemicals, sodium hydroxide and sodium sulfide
Cogenerating steam and power
Source: *Based on Tran and Vakkilainen (2007).*

The importance of the recovery process is often not fully valued. Over 1.3 billion tons per year of weak black liquor are processed globally; about 200 million tons per year of black liquor dry solids are burned in recovery boilers to recover 50 million tons of cooking chemicals as sodium oxide (Na_2O) and to produce 700 million tons of high-pressure steam. This makes black liquor the fifth most important fuel in the world, next to coal, oil, natural gas, and gasoline (Reeve, 2002). Since black liquor is derived from wood, it is the most important renewable biofuel, particularly in Sweden and Finland. The chemical, physical, and combustion properties of black liquor vary from mill to mill depending on many factors, including mill location (inland or coastal), digester conditions, pulp yield, wood species, white liquor properties, chemicals-to-wood ratio, and brownstock washing efficiency. In general, hardwood pulping requires less chemicals, has a higher pulp yield, and consequently, generates less BLS than softwood pulping. Harwood black liquor generally contains less organics, tall oil and soap, and has a lower heating value (about 5% lower) than softwood black liquor. In Brazil, Chile, and tropical countries, eucalyptus is the dominant wood species used in kraft pulping. Since the properties of eucalyptus black liquor are similar to those of other types of hardwood black liquor, the chemical recovery process in eucalyptus kraft mills is essentially the same as others.

The black liquor, includes minerals of white liquor spent during cooking and dissolved organic substances from wood (Grigoray, 2009). The liquor is an important by-product since its combustion results in not only chemical recovery but also a significant amount of heat, which covers the production costs of energy. The spent liquor segregated during pulp washing contains a large quantity of water that is removed by evaporation. The black liquor dry residue consists of 30–40% minerals and 60–70% of organic matters. Of the mineral residue, 18–25% of it is chemically connected with organic substances of dissolved wood, 1–2% of the mineral residue is in the form of free alkali, 1–4% in the form of sodium sulfide, 3–5% in the form of sodium sulfate, and 4–10% in the form of sodium carbonate. The organic part of the dry residue is made up of lignin (30–35%) and products of carbohydrates destruction (30–35%). The composition of this combustible mass includes 35–45% carbon, 3–5% hydrogen, 15–20% oxygen, and 1–4% organic sulfur (Grigoray, 2009). Elemental composition of the liquor is determined by the type of wood and conditions of its delignification. Typical elemental analysis of BLS is shown in Table 1.7.

The physical properties of black liquor are directly dependent on its composition. The liquor properties which describe its behavior during the heat transfer processes are of interest to study (Fakhrai, 1999). The main physical parameters of sulfate black liquor considered during its evaporation and burning are density, viscosity, boiling temperature, surface tension, and heat value (Adams, 1989). The liquor density characterizes the dry solids content. During the rise in temperature, density decreases due to volumetric expansion of water contained in the liquor (Bogomolov, 1989). The viscosity of spent liquor depends on the amount of dry solids, their chemical composition, and liquor temperature. An increase in the concentration of dry solids leads to increased viscosity, which in turn contributes to lower cost energy when burning liquor. On the other hand, a very high concentration of black liquor requires using medium steam pressure in the evaporation process, which demands additional energy consumption. Also the viscosity of black liquor is limited by the throughput of the pumps used

Table 1.7 Typical Elemental Analysis of BLS	
Moisture (wt% as received)	27.9
C (wt% dry)	32.5
Na (wt% dry)	20.4
S (wt% dry)	5.9
H (wt% dry)	3.4
K (wt% dry)	2.16
Cl (wt% dry)	0.14
N (wt% dry)	0.08
Si (mg/kg dry)	89
Mg (mg/kg dry)	83
Ca (mg/kg dry)	78
Mn (mg/kg dry)	40
Fe (mg/kg dry)	19
Al (mg/kg dry)	12
Ni	n.a.
Br	n.a.
F	n.a.
I	n.a.
Lower heating value (MJ/kg dry)	12.0
Source: *Based on Nordin (1994) and Råberg (2007a,b).*	

for liquor transfer. An effective way to reduce viscosity is to keep the liquor under a temperature of 180°C for 30 min since this destroys long organic molecules (Holmlund and Parviainen, 1999). The boiling temperature of black liquor is higher than the boiling temperature of water at the same pressure and varies with the composition of dissolved organic substances. This difference in temperature is called the boiling point rise. This parameter is the basis of the evaporation process. The problems of foaming in the evaporation process are associated with the low value of surface tension of black liquor. Surface tension decreases with the increasing temperature and the decreasing concentration of the solution. The presence of turpentine and soap decreases surface tension, therefore, they should be removed from the liquor (Holmlund and Parviainen, 1999). The heat value is one of the most important parameters of black liquor because it shows the quantity of heat which can be obtained by burning. Organic and inorganic components of the liquor have different heat values as indicated in Table 1.8 (Henricson, 2004). Reactions taking place during burning

Table 1 8 Heat Values of Spent Liquor Components		
Component	MJ/kg	Btu/lb
Softwood lignin	26.90	11.57
Hardwood lignin	25.11	10.80
Carbohydrates	13.56	5.83
Resins, fatty acids	37.71	16.22
Sodium sulfide	12.90	5.55
Sodium thiosulfate	5.79	2.49
Source: *Based on Henricson (2004).*		

also affect the liquor's heat value, e.g., the recovery reaction of sodium sulfate to sulfide consumes energy.

The black liquor is transformed to white liquor via a sequence of steps. First, black liquor is sent through a series of evaporators to increase the solid content in order to make it suitable for combustion. In the evaporators, the BLS content is increased from about 15 wt% to about 70–75 wt%. The concentrated black liquor that goes into the recovery boiler consists of about 30% water, 30% valuable inorganic cooking chemicals, and 40% lignin and other organic matter separated from the wood. The higher heating value is about 14 MJ/kg solids. Typically, a large pulp mill will produce about 2000 ton per day of black liquor corresponding to about 300 MWth. It is important to recover both chemicals and energy from the black liquor for economical and environmental purposes. In a conventional kraft pulp mill, the energy in the black liquor is recovered as steam by combustion in a recovery boiler. In the recovery boiler, chemicals are recovered as a smelt that is converted to new white liquor. Most of the recovery boilers used today is of the Tomlinson type (Adams, 1997). A recovery boiler acts both as a type of high-pressure steam boiler and as a chemical reactor with reductive and oxidative zones. In the recovery boiler, black liquor is sprayed with a splash plate nozzle into the boiler at a temperature around 120°C. The drying and pyrolysis conversion stage occur quickly before most droplets fall onto a bed of char where the final conversion takes place. The resulting inorganic smelt exits via a channel through the wall and dissolves into a liquid solution to form green liquor. The combustible gases released during the pyrolysis are completely burned by different air registers along the main wall. Then steam is produced from the combustion heat in surrounding water pipes of the boiler, superheaters, and economizer. The Tomlinson boiler has served the pulping industry for more than 70 years, mainly because of its high availability and stable running. Also continuous efforts have been made to increase its efficiency, but still it has a relative low overall efficiency for generation of electricity. The black liquor droplets being sprayed into the recovery boiler are undergoing the following three stages: drying, pyrolysis, and char conversion (Hupa, et al., 1987). First the droplet dries and evaporates its moisture. Then the droplet is pyrolyzed and the organic matter in the liquor is released as various gaseous compounds from the volatile substances. The rate of drying and pyrolysis is controlled by the heat transfer rate to the droplet. The resulting pyrolysis gases are mainly H_2, CO, CO_2, H_2O, and some heavier hydrocarbons (Bhattacharya et al., 1986). The remaining droplet is a swollen porous char particle containing mainly carbon and sodium salts. The last stage is char conversion, in which gas phase species react with the organic content in the char particle and convert it into gaseous species. At the end of this stage, the droplet would ideally consist only of inorganic salts, primarily sodium carbonate (Whitty et al., 1998).

The black liquor burned in the recovery furnace has a high energy content (5800–6600 British thermal units per pound [Btu/lb] of dry solids), which is recovered as steam for process requirements, such as cooking wood chips, heating and evaporating black liquor, preheating combustion air, and drying the pulp or paper products. The process steam from the recovery furnace is often supplemented with fossil fuel-fired and/or wood-fired power boilers. Particulate matter (PM) (primarily sodium sulfate (Na_2SO_4)) exiting the furnace with the hot flue gases is collected in an electrostatic precipitator (ESP) and added to the black liquor to be fired in the recovery furnace. Additional makeup sodium sulfate, or "saltcake," may also be added to the black liquor prior to firing. Molten inorganic salts, referred to as "smelt," collect in a char bed at the bottom of the furnace. Smelt is drawn off and dissolved in weak wash water in the smelt dissolving tank (SDT) to form a solution of carbonate salts called "green liquor," which is primarily sodium sulfide and sodium carbonate (Na_2CO_3). Green liquor also contains insoluble unburned carbon and inorganic impurities, called dregs, which are removed in a series of clarification tanks. The green liquor then goes through a causticizing process to convert green liquor back to white liquor. Decanted green liquor is transferred to the causticizing area, where sodium carbonate is converted to sodium hydroxide by the addition of lime (calcium oxide

(CaO)). The green liquor is first transferred to a slaker tank, where calcium oxide from the lime kiln reacts with water to form calcium hydroxide (Ca(OH)$_2$). From the slaker, liquor flows through a series of agitated tanks, referred to as causticizers, that allow the causticizing reaction to go to completion (i.e., calcium hydroxide reacts with sodium carbonate to form sodium hydroxide and calcium carbonate (CaCO$_3$)). The causticizing product is then routed to the white liquor clarifier, which removes calcium carbonate precipitate, referred to as "lime mud." The lime mud is washed in the mud washer to remove the last traces of sodium. The mud from the mud washer is then dried and calcined in a lime kiln to produce "reburned" lime, which is reintroduced to the slaker. The mud washer filtrate, known as weak wash, is used in the SDT to dissolve recovery furnace smelt. The white liquor (sodium hydroxide and sodium sulfide) from the clarifier is recycled to the digesters in the pulping area of the mill.

1.4 MODERN KRAFT MILLS

The design of a modern twenty-first century pulping complex is dramatically different from mills of the past. Pressures to maximize energy efficiency, improve product quality, reduce environmental impact, and optimize capital and operating costs have significantly shaped twenty-first century pulping and bleaching processes. New mills have responded to these demands by adopting efficient, low impact designs on economies of scale that far surpass most existing mills. New fiber lines have been built mainly in Asia and South America where access to fast growing raw material and other production advantages gives favorable levels of cost and return. A modern kraft pulp mill is energy self-sufficient; it can produce all the steam and power that is needed for the process as well as a surplus. The major part of the energy comes from the combustion of black liquor in the recovery boiler. The second boiler on-site (power boiler in Figure 1.3) is used to burn the bark and sometimes biosludge from the effluent treatment. In older, less energy-efficient, pulp mills and in most integrated mills, purchased fuels are also burned in the power boiler. These are mostly wood fuels, but some oil is also used. The lime kiln is then usually fired with oil or natural gas, but in modern market pulp mills the surplus of biofuels is used to provide heat also for the lime kiln. Available methods include bark gasification and direct-firing of pulverized bark. High-pressure steam is generated in both boilers, and electric power is generated in one or two back pressure steam turbines. The heat demand of the mill is usually split between two steam levels, medium pressure (MP) at 10–12 bar and low pressure (LP) at 4–5 bar. The logistics of handling biomass feedstock are well developed around a pulp mill. Large mills that produce in the order of 2000 tons of pulp per day handle 3–4 million cubic meters of wood per year. In energy terms, the wood that is processed corresponds to 800–900 MW.

Weak black liquor is produced at 15% solids, which is then evaporated to 78% solids using a seven-effect multiple-effect evaporation system with an integrated superconcentrator. Steam economy is estimated at 6.0 kg water/kg steam for such an arrangement. The black liquor contains 1.6 kg solids/kg pulp. The steam requirement for the evaporators is calculated to be 3.1 GJ/ADt. The electricity requirement would be 30.0 GJ/ADt (NRC, 2009). The mill would use a high solid recovery boiler that achieves a 75% heat-to-steam efficiency. Heating value of black liquor would be approximately 6250 Btu/lb solids (21.0 GJ/ADt pulp). The boiler air is heated to 150°C using steam, and minimal use of soot blowers is employed, consuming 0.9 GJ/ADt of steam. The mill would recover 70% of steam condensate, and the resulting energy use in the deaerator is 1.0 GJ/ADt. Condensing–extracting steam turbines are used to produce electricity with a power-to-heat ratio of 100 kWh/GJ. The mills power boiler uses hog fuel and achieves a heat-to-steam efficiency of 70%. The mill generates 15.8 GJ/ADt in the recovery boiler to satisfy the heat requirements of the process and electricity generation needs. A back-pressure steam turbine generates 520 kWh/ADt of electricity, while excess high-pressure steam generates another 135 kWh/ADt through a steam-condensing turbine. Therefore, the total electricity generation by the mill is 655 kWh/ADt. The power plant has a parasitic electricity need of about 60 kWh/ADt.

State-of-the-art cooking includes both continuous and batch processes utilizing low cooking temperatures and optimized alkali profiles (Johnson et al., 2008). Continuous cooking has predominated over the last decade and typically consists of two vessels for softwood and one or two vessels for hardwood. The chips are segregated by species. Thus the chip feed is of uniform quality resulting in reduced processing upsets. Lower cooking temperatures (145–153°C) are typically used with kappa numbers in the range 26–35 for softwoods, 17–22 for birch, and 15–18 for eucalypts.

Atmospheric diffusion washers and drum washers common in twentieth century mills are replaced by presses and multi-stage Drum displacer (DD) washers that achieve high equivalent lent displacement ratios (EDRs). Oxygen delignification is carried out at medium consistency and in two stages by all three mills. This is a significant difference to mills of the previous decade where oxygen delignification was typically one stage. The bulk of the delignification occurs in the first reactor, which is typically run at lower temperature and higher pressure than the second reactor. Veracel runs its first reactor at a temperature of 92–96°C, a pressure of 6–8 bar, with the addition of 60–70% of the alkali and oxygen charge. Valdivia adds all chemicals to the first reactor. Typically, the retention time in the first tower is about half that of the second tower. Veracel operates the second reactor at a temperature of 98–100°C and a pressure of 3–5 bar. Oxygen delignification at the Hainan Jinhai mill is the two-stage Dualox™ process. Bleaching sequences used are mostly Elemental chlorine free (ECF).

REFERENCES

Ådahl, A., Harvey, S., Berntsson, T., 2004. Process industry energy retrofits: the importance of emission baselines for greenhouse gas reductions. Energy Policy 32, 1375–1388.

Adams, T., 1989. In: Grace, T., Malcolm, B. (Eds.), Pulp and Paper Manufacture, Alkaline Pulping, third ed. TAPPI, Georgia, USA, p. 637.

Adams, T.N., 1992. Black liquor combustion. In: third ed. Kocurek, M.J. (Ed.), Pulp and Paper Manufacture, vol. 5. Joint Committee of TAPPI and CPPA, Atlanta, GA, p. 531.

Adams, T.N., 1997. General characteristics of kraft black liquor recovery boilers. In: Adams, T. N. (Ed.), Kraft Recovery Boilers 6B. TAPPI Press, Norcross, GA, p. 3. (Chapter 1).

Andersson, E., Harvey, S., 2004. Pulp-mill integrated bio-refineries: a framework for assessing net CO_2 emission consequences. In: Proceedings, AIChE 2004 Fall Annual Meeting, Austin, TX, pp. 203–208.

Bajpai, P., 2008. Chemical Recovery in Pulp and Paper Making. PIRA International, UK, 166pp.

Bajpai, P., 2012. Biotechnology in Pulp and Paper Processing. Springer-Verlag, New York, NY.

Berglin, N., Eriksson, H., Berntsson, T., 1999. Performance evaluation of competing designs for efficient cogeneration from black liquor. In: Second Biennial Johan Gullichsen Colloquium, Helsinki, Finland.

Berglin, N., Lindblom, M., Ekbom, T., 2002. Efficient production of methanol from biomass via black liquor gasification. In: Tappi Engineering Conference, San Diego, CA.

Berglin, N., Lindblom, M., Ekbom, T., 2003. Preliminary economics of black liquor gasification with motor fuels production. In: Colloquium on Black Liquor Combustion and Gasification, Park City, UT, May 13–16, 2003.

Bhattacharya, P.K., Parthiban, V., Kunzru, D., 1986. Pyrolysis of black liquor solids. Ind. Eng. Chem. Proc. Design Dev. 25 (2), 420–426.

Biermann, C.J., 1996a. Pulping fundamentals. Handbook of Pulping and Papermaking. Academic Press, San Diego, CA, p. 55.

Biermann, C.J., 1996b. Kraft spent liquor recovery. Handbook of Pulping and Papermaking. Academic Press, San Diego, CA, p. 101.

Bogomolov, B.D., 1989. The Conversion of Sulfate and Sulphite Liquors. Forest Industry, Moscow, pp. 23–24.

Dahlquist, E., 2003. A combined physical and statistical simulation model for black liquor gasification. In: SIMS 2003 Conference in Vasteras.

Dahlquist, E., Jacobs, R., 1994. Development of a dry black liquor gasification process. Pulp Pap. Canada 95, 2.

Dahlquist, E., Jones, A., 2005. Presentation of a dry black liquor gasification process with direct caustization. TAPPI J. 15, 19.

Dahlquist, E., Avelin, A., Yan, J., 2009. Black liquor gasification in a CFB gasifier—system solutions. In: The First International Conference on Applied Energy (ICAE'09), Hong Kong, January 5–7, 2009.

Ekbom, T., Lindblom, M., Berglin, N., Ahlvik, P., 2003. Technical and commercial feasibility study of black liquor gasification with methanol/DME production as motor fuels for automotive uses—BLGMF. Report for Contract No. 4.1030/Z/01-087/2001, European Commission, Altener program, Stockholm, Sweden.

Eriksson, H., Harvey, S., 2004. Black liquor gasification—consequences for both industry and society. Energy 29, 581–612.

Fakhrai, R., 1999. Modelling of Carry-over in Recovery Furnace (Licentiate thesis). Royal Institute of Technology, Department of Metallurgy.

Farmer, M.C., Sinquefield, S., 2009. Black Liquor Gasification: Paths and Obstacles to Commercialisation. Pira Industry Insight, UK.

Grigoray, O., 2009. Gasification of Black Liquor as a Way to Increase Power Production at Kraft Pulp Mills (Master thesis). Lappeenranta University of Technology Faculty of Technology Degree Program of Chemical Technology.

Gullichsen, J., 2000. Fiber line operations. In: Gullichsen, J., Fogelholm, C.-J. (Eds.), Chemical Pulping—Papermaking Science and Technology, Book 6A. Fapet Oy, Helsinki, Finland, p. A19.

Harvey, S., Facchini, B., 2004. Predicting black liquor gasification combined cycle powerhouse performance accounting for off-design gas turbine operation. Appl. Therm. Eng. 24, 111–126.

Henricson, K., 2004. An introduction to chemical pulping technology. Lecture Notes, 2004.

Holmberg, J., Gustavsson, L., 2007. Chemical mechanical biomass use in chemical and mechanical pulping with biomass-based energy supply. Resour. Conserv. Recycling 52, 331–350.

Holmlund, K., Parviainen, K., 1999. Evaporation of black liquor, chemical pulping. In: Gullichsen, J., Fogelholm, C. (Eds.), Papermaking Science and Technology, Book 6B. Fapet Oy, Helsinki, Finland, pp. 37–65.

Hupa, M., Solin, P., Hyoty, P., 1987. Combustion behaviour of black liquor droplets. J. Pulp Paper Sci. 13 (2), 67–72.

IEA Bioenergy Report, 2009. Black Liquor Gasification Summary and Conclusions from the IEA Bioenergy ExCo54 Workshop.

Joelsson, J., Gustavsson, L., 2008. CO_2 emission and oil use reduction through black liquor gasification and energy efficiency in pulp and paper industry. Resour. Conserv. Recycling 52, 747–763.

Johnson, A.P., Johnson, B.I., Gleadow, P., Silva, F.A., Aquilar, R.M., Hsiang, C.J., et al., 2008. 21st century fibre lines. In: Proceedings of International Pulp Bleaching Conference, Quebec City, p. 1.

Jonsson, M., Yan, J., 2005. Humidified gas turbine—a review of proposed and implemented cycles. Energy 30, 1013–1078.

Kassberg, M., 1998. Lutförbränning sulfat och sulfit, Skogsindustrins utbildning i Markaryd AB.

Larson, E.D., Consonni, S., Katofsky R.E., 2003. A cost-benefit assessment of biomass gasification power generation in the pulp and paper industry, Princeton Environmental Institute 2003. Available at <www.princeton.edu/~energy>.

Maunsbach, K., Isaksson, A., Yan, J., Svedberg, G., Eidensten, L., 2001. Integration of advanced gas turbines in pulp and paper mills for increased power generation. J. Eng. Gas Turbines Power Trans. ASME 123 (4), 734–741.

Möllersten, K., Yan, J., Westermark, M., 2003a. Potential and cost-effectiveness of CO_2-reduction in the Swedish pulp and paper sector. Energy 28, 691–710.

Möllersten, K., Yan, J., Moreira, J., 2003b. Potential market niches for biomass energy with CO_2 capture and storage. Biomass Bioenergy 25 (3), 273–285.

Möllersten, K., Lin, G., Yan, J., Obersteiner, M., 2004. Efficient energy systems with CO_2 capture and storage from renewable biomass in pulp and paper mills. Renew. Energy 29, 1583–1598.

NRC, 2009. The model kraft market pulp mill. <www.nrcan.gc.ca>.

Naqvi, M., Yan, J., Dahlquist, E., 2010. Black liquor gasification integrated in pulp and paper mills: A critical review. Bioresour. Technol. 101, 8001–8015.

Näsholm, A., Westermark, M., 1997. Energy studies of different cogeneration systems for black liquor gasification. Energy Convers. Manage. 15 (38), 1655–1663.

Nordin, A., 1994. Chemical elemental characterization of biomass fuels. Biomass Bioenergy 6, 339–347.

Patrick, K., 2003. Gasification edges closer to commercial reality with three new N.A. Mill start-ups. PaperAge, October 2003, pp. 30–33

Råberg, M., 2007a. Black Liquor Gasification—Experimental Stability Studies of Smelt Components and Refractory Lining. Energy Technology and Thermal Process Chemistry (Ph.D. thesis). Umeå University, Umeå.

Råberg, M., 2007b. Black Liquor Gasification. Division of Energy Engineering, Department of Engineering Sciences and Mathematics, Luleå University of Technology, Luleå, Sweden.

Reeve, D.W., 2002. The kraft recovery cycle. TAPPI Kraft Recovery Operations Short Course. TAPPI Press, Norcross, GA.

Smook, G.A., 1992. Overview of pulping methodology, Handbook for Pulp and Paper Technologists, second ed. Angus Wilde Publications, Vancouver, p. 36.

Sricharoenchaikul, V., 2009. Assessment of black liquor gasification in supercritical water. Bioresour. Technol. 100, 638–643.

Swedish Forest Agency, 2008. Forestry Statistics. Swedish Forest Agency, Jönköping, Sweden, <www.svo.se>.

Tran, H.N., Vakkilainen, E.K., 2007. Advances in the kraft chemical recovery process. In: Third ICEP International Colloquium on Eucalyptus Pulp, Belo Horizonte, Brazil, March 4–7.

U.S. Energy Information Administration, 2009. Independent Statistics and Analysis. U.S. Department of Energy, Washington, DC, <http://www.eia.doe.gov/oiaf/ieo/world.html>.

Vakkilainen, E.K., 2000. Chemical recovery. In: Gullichsen, J., Paulapuro, H. (Eds.), Papermaking Science and Technology, Book 6B. Fapet Oy, Helsinki, Finland, p. 7. (Chapter 1).

Waldner, M., Vogel, F., 2005. Renewable production of methane from woody biomass by catalytic hydrothermal gasification. Ind. Eng. Chem. Res. 44 (13), 4543–4551.

Whitty, K., Backman, R., Hupa, M., 1998. Influence of char formation conditions on pressurized black liquor gasification rates. Carbon 36 (11), 1683–1692.

Yan, J., Dahlquist, E., Jin, H., Gao, L., Tu, S., 2007. Integration of large scale pulp and paper mills with CO_2 mitigation technologies. In: The Third International Green Energy Conference, Västerås, Sweden.

Yan, J., Eidensten, L., 2000. Status and perspective of externally fired gas turbines. J. Propul. Power 16, 572–576.

Yan, J., Eidensten, L., Svedberg, G., 1995. An investigation of heat recovery system in externally fired evaporative gas turbines. ASME Paper 95-GT-72.

Black Liquor Gasification

Black liquor is a biomass feedstock with unique properties suitable for gasification (Andersson and Harvey, 2004; Ådahl et al., 2004; Berglin et al., 2002; Dahlquist, 2003; Dahlquist and Jones, 2005; Dahlquist et al., 2009; Ekbom et al., 2003; Harvey and Facchini, 2004; Möllersten et al., 2003a,b, 2004; Maunsbach et al., 2001; Waldner and Vogel, 2005; Sricharoenchaikul, 2009; Bajpai, 2008, 2013; Dance, 2005; Marklund, 2006; Grigoray, 2009; Landälv, 2010; Salomonsson,

2013). First of all, it is available at existing industrial sites in large quantities. Second, it is a liquid. This makes it possibly to easily feed it by pumping into the pressurized gasifier. With biomass in solid or pulverized form this becomes significantly more difficult. The liquid state also makes the black liquor easy to atomize into a fine mist that reacts very fast in the gasifier. Third, the gasification of black liquor char is more rapid than for any other feedstock as the inherently high sodium and potassium content of black liquor acts as a catalyst. These properties make it possible to apply the high temperature, entrained flow gasification principle to black liquor. This type of gasification process provides many advantages over alternative gasification technologies: it is a very rapid, single-stage gasification process with low reactor volume; it minimizes the need for raw syngas cleanup as the Chemrec process directly provides a raw syngas of excellent quality with complete carbon conversion; no tar formation; and low methane content.

These properties make the gasification of black liquor easier and more rapid than for any other biomass feedstock. BLG should be an integral part of an Integrated Forest Products Biorefinery (IFBR) because its process heat may be used in the sugar conversion unit operations, and the synthesis gas may be used to replace fossil fuels, in particular oil in the lime kiln. The gasification synthesis gas may be used as feedstock to produce transportation fluids such as Fisher—Tropsch (FT) liquid hydrocarbons, methanol, and mixtures of higher alcohols (Bajpai, 2013). The key requirement for implementation of BLG is to demonstrate the reliability and efficiency of the technology at commercial scale while regenerating the pulping chemicals.

BLG is one of several biorefinery options for the kraft pulp industry. It enables production of several value-added products such as electricity, district heating, biofuels, or lignin in addition to pulp. Investment in biorefineries is a possible way for the industry to remain competitive with increased energy and raw material prices. Some mills, especially energy-efficient market kraft pulp mills, have the possibility to become major net exporters of electricity or lignin without purchasing external wood fuel (Pettersson, 2011). Nevertheless, in order for integrated pulp and paper mills, even those with a high degree of energy efficiency, to become major exporters of lignin or for any type of mill to become a major exporter of biofuels, external wood fuel is required. In such cases, the usage of biomass should be compared with other possible ways to use the biomass resource. Increasing the degree of heat integration could decrease the need for external process heating and thereby decrease the need for external wood fuel. If the technology becomes commercially available, implementation of carbon capture and storage (CCS) could significantly influence both the climate impact and economic performance of the studied systems. In BLG, and other gasification processes, relatively large amounts of CO_2 could be captured at relatively low costs.

Large amounts of CO_2 could be separated from the flue gases of the recovery boiler or other mill power boilers. But the separation costs are generally very much higher compared to implementation in the gasification processes. In order for mills to consider implementation of full-scale BLG plants, the recovery boiler has to be close to the end of its technical lifetime. But mills with a steam surplus, or mills planning to increase their production capacity (assuming that the recovery boiler is running at maximum capacity), could consider investment in a smaller BLG plant as a way to take advantage of a potential steam surplus or to achieve debottlenecking of the recovery boiler.

BLG is being considered primarily as an option for production of biofuels in recent years due to the focus on the transport sector's high oil dependence and climate impact (Pettersson, 2011). The technology is included in several studies comparing climate and economic benefits of alternative ways to produce motor fuels. But there is a tendency to present BLG without consideration of the special implementation characteristics of each specific case. Some studies discuss how other characteristics, such as integration with another type of mill, would affect the results.

According to Pettersson (2011), estimating the climate impact and economic performance of possible future technologies is not straight-forward. Uncertainties about future energy prices and policy instruments make the results highly variable. Also, when it comes to estimation of climate impact, a number of different approaches can be considered with significant variation of results. Gasification of black liquor is an alternative recovery technology that has gone through a stepwise development since its early predecessor was developed in the 1960s. The currently most commercially advanced BLG technology is the Chemrec® technology (Chemrec, 2013), which is based on entrained flow gasification of the black liquor at temperatures above the melting point of the inorganic chemicals. In a BLG system, the recovery boiler is replaced with a gasification plant. The evaporated black liquor is gasified in a pressurized reactor under reducing conditions. The generated gas is separated from the inorganic smelt and ash. The gas and smelt are cooled and separated in the quench zone below the gasifier. The smelt falls into the quench bath where it dissolves to form green liquor in a manner similar to the dissolving tank of a recovery boiler. The raw fuel gas exits the quench and is further cooled in a countercurrent condenser (CCC). Water vapor in the fuel gas is condensed, and this heat release is used to generate steam. Hydrogen sulfide is removed from the cool, dry fuel gas in a pressurized absorption stage. The resulting gas is a nearly sulfur-free synthesis gas (*syngas*) consisting of mostly carbon monoxide, hydrogen, and carbon dioxide.

The two main technologies under development are pressurized gasification and atmospheric gasification, being commercialized by Chemrec AB and ThermoChem Recovery International (TRI), respectively (Chemrec, 2013; TRI, 2013).

2.1 BLG TECHNOLOGIES

BLG may be performed either at low temperatures or at high temperatures based on whether the process is conducted above or below the melting temperature range (650–800°C) of the spent pulping chemicals (Sricharoenchaikul, 2001).

Low temperature gasification —Low temperature gasifier operates at 600–850°C, below the melting point of inorganics, thus avoiding smelt–water explosions.

High temperature gasification—High temperature gasification units generally operate in the 900–1000°C range and produce a molten smelt.

In low temperature gasification, the alkali salts in the condensed phase remain as solid products while molten salts are produced in high temperature gasification. Low temperature gasification is advantageous over high temperature gasification because gasification at low temperatures yields improved sodium and sulfur separation. Additionally, low temperature gasification requires fewer constraints for materials of construction because of the solid product. However, the syngas of low temperature gasification may contain larger amounts of tars, which can contaminate gas cleanup operations in addition to contaminating gas turbines upstream of the gasifier. These contamination problems can result in a loss of fuel product from the gasifier (Sricharoenchaikul, 2001; Patrick and Siedel, 2003).

Several companies have performed trials to develop a commercially feasible process for BLG. History of BLG development is very well described by Whitty and Baxter (2001) and Whitty and Verrill (2004). In the search for alternative ways of recovering the cooking chemicals, gasification techniques have been thoroughly examined several times. In total, more than 20 different technologies have been investigated over the years (Swedish Energy Agency, 2008). The most interesting attempts of accomplishing low temperature black liquor gasification (LTBLG) and high temperature black liquor gasification (HTBLG) processes are presented in Table 2.1.

Only two technologies are currently being commercially pursued: the MTCI (Manufacturing and Technology Conversion International Company, low temperature) and Chemrec (high temperature) technologies. Table 2.2 shows the difference between these two technologies.

Weyerhaeuser, New Bern, uses a Chemrec booster for BLG but it operates at atmospheric pressure, which does not give maximum energy efficiency. Energy efficiency is enhanced by going to higher pressures. Trials were run at Kappa Kraftliner, Sweden, in which the black liquor was gasified at high temperature and pressure in a reactor then the gas was cooled and separated from droplets of smelt. The condensate was dissolved to form low-sulfidity green liquor. The raw gas containing carbon monoxide and carbon dioxide was saturated with steam at high pressure then cooled and stripped of particles. The gas can be used as a feedstock in a combined cycle technology or for chemical synthesis (Larson et al., 2000).

The development of various BLG technologies is discussed below (Whitty and Verrill, 2004; Whitty, 2005; Whitty and Baxter, 2001; Naqvi et al., 2010; Empie, 1991; Grigoray, 2009).

Table 2.1 BLG Processes

SCA-Billerud process
Copeland process
Weyerhaeuser's "dry pyrolysis" process
St. Regis hydropyrolysis process
The Texaco coking process
VTT' circulating fluidized gasification process
Babcock & Wilcox's bubbling fluidized bed gasification process
NSP or Ny Sodahus Process
DARS process (Direct Alkali Recovery System)
ABB circulating fluidized bed gasification process
MTCI/TRI's fluidized bed steam reforming process
Kellogg, Brown, and Root's spouting and transport fluidized bed gasifiers
Paprican's AST process
University of California's "pyrolysis, gasification, combustion" process
The Champion–Rockwell molten salt gasification process
The SKF plasma black liquor gasifier
Ahlstrom's suspension gasifier
Tampella's entrained flow gasifier
Noell's entrained flow gasifier
Champion/Rockwell molten salt gasification process
Chemrec Entrained flow gasification process
Catalytic hydrothermal gasification of black liquor

Source: *Based on Whitty and Verrill (2004), Grigoray (2009), Naqvi et al. (2010), Swedish Energy Agency (2008).*

Table 2.2 Difference Between LTBLG and HTBLG

Property	LT (TRI)	HT (Chemrec)
Heating	Indirectly (syngas)	Directly (black liquor)
Chemicals recovered	Solid phase	Smelt
Sulfur split (gas/smelt)	90/10	50/50
Syngas composition	High H_2-concentration	Moderate H_2-concentration
Syngas energy content	High	Low

Source: *Based on Swedish Energy Agency (2008).*

2.1.1 SCA-Billerud Pyrolysis Process

This technology was operated from 1958 to 1980 (Horntvedt, 1968, 1976; Maloney, 1976). But it has been abandoned because of the technical reasons. The process was developed by SCA-Billerud and was implemented at Ortviken (Dahlquist and Jacobs, 1994; Whitty and Verrill,

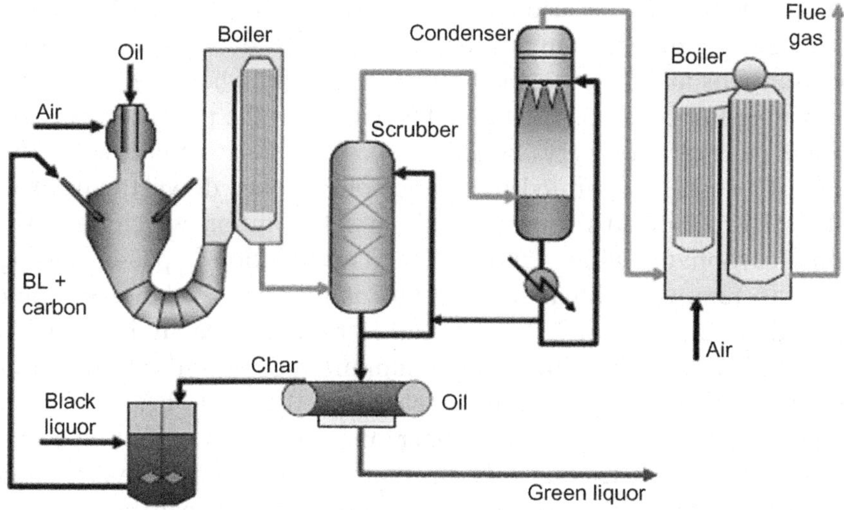

Figure 2.1 SCA-Billerud pyrolysis process. Reproduced with permission from Whitty and Verrill (2004), adaptation of Maloney (1976).

2004). Development of the SCA-Billerud process began in the late 1950s. The first commercial plant was built in Sundsvall, Sweden, in 1968 which was successful. Later more commercial plants were built. Most of the commercial plants were running sulfite or Neutral Sulfite Semi Chemical (NSSC) liquor. Substantial efforts were made into modifying the SCA-Billerud pyrolysis process for kraft liquor. A modified plant of 25 ton per day was started up at a kraft mill in Pamplona, Spain, in 1970 to evaluate the process on black liquor from kraft process and also to provide incremental capacity. Later advanced configurations for kraft liquor recovery were suggested. However, these were not studied on commercial scale. Figure 2.1 shows SCA-Billerud pyrolysis process. The pyrolyzer is a downward flowing entrained flow reactor and is cone shaped. At the top, it has an oil burner firing axially through the reactor. Nozzles are located around the periphery of the reactor near the top. These inject black liquor and recycled carbon into the hot reaction zone. Here the liquor dries and undergoes pyrolysis. The reactor is run under substoichiometric conditions and at temperatures below the melting point of the ash (Whitty and Verrill, 2004).

The outlet of the reactor goes into a tail of piping that turns from a downward flow to an upward flow. Most of the ash remains entrained in the gas stream, but some ash deposits on the walls of the reactor and tail. It must be occasionally removed. This is attained by running the reactor sometimes in an oxidizing mode at high temperature. The ash melts and flows to the bottom of the tail where it falls through an opening into a dissolving tank. The gas from the pyrolyzer is passed through a heat recovery boiler to provide relatively low-pressure steam. The ash is separated in a series of cyclones downstream of the boiler. It is dissolved to form a sodium salt solution which comprises mainly carbonate, sulfate, and sulfide. This solution is passed through a filter to remove residual carbon. This is recycled to the black liquor supplied to the reactor. The dry gas is particle free. It is fired in a power boiler, usually with oil as a support fuel. For the kraft liquor, the process is modified to feed the gas from the waste heat boiler into a combined wet particulate scrubber and hydrogen sulfide absorber. The liquid product leaving this unit contains carbonate, hydrosulfide, and carbon. After filtering, it is green liquor which can be processed by conventional means.

The main problems of this process were a low carbon conversion attributable to a very short residence time. Still the process operated with high availability, and less manpower was needed for the conventional recovery boiler that was running in parallel (Whitty, 2005; Whitty and Verrill, 2004; Grigoray, 2009). Five commercial plants were installed in Sweden, Japan, and the United States, serving mills ranging in size from 270 to 1000 tons pulp per day.

2.1.2 The Copeland Recovery Process

The Copeland recovery process was based on dense bubbling fluidized bed comprising of black liquor ash (sodium carbonate and sulfate) and optional inert material which was silica sand (Whitty and Verrill, 2004; Copeland and Hanway, 1967; Copeland, 1969; Evans, 1975). George Copeland brought fluidized bed technology to the pulp and paper industry during the 1960s and 1970s. He was the founder of Copeland Systems, Inc. Most of the Copeland recovery systems were initially installed on sulfite and soda mills having low concentrations of chlorine in the black liquor. This helped in maintaining a higher ash melting temperature which reduced the risk of bed agglomeration. Later, the Copeland process was applied to black liquor generated from the Kraft process. The first installation was in a Canadian mill. This mill processed 300 metric tons of dry solids per day black liquor. The major use of the Copeland process was to provide incremental recovery capacity for overloaded recovery boilers. Dorr-Oliver developed a fluidized bed liquor combustion technology which was very similar to the Copeland process. Several reasons contributed to the closure of this technology. An increase in nonprocess elements of the black liquor which resulted from process closure reduced the melting point of the liquor ash. This increased the risk of bed agglomeration. This technology became uncompetitive due to improvements made in operation and safety of the conventional recovery boiler technology and push towards higher energy efficiency and higher solids content in the black liquor (Whitty and Verrill, 2004).

Figure 2.2 shows the Copeland recovery process (Copeland and Hanway, 1967). Black liquor of low solid content is sprayed on the surface of the bed from above the freeboard region. The bed was fluidized by using the preheated air. The air was fed at such a rate to keep combustion stoichiometric. As a result, the oxygen content in the freeboard was near to zero. Bed temperatures were kept below the melting temperature of the ash by regulating the amount of preheat on the incoming air and by feeding liquor of low solids. Afterwards, the system was studied in reducing mode, i.e., gasification, but the process was not commercialized. Bed material was constantly removed from the bottom of the bed, and either dissolved and fed to the causticizing system. For kraft mills, solids were fed to the recovery boiler to reduce the sulfate. Later, Copeland developed a separate fluidized bed for reduction of the solids. Flue gas from the process was passed through one or more cyclones to remove particulates, passed through a venturi scrubber and exhausted to the atmosphere.

2.1.3 Weyerhaeuser's Process

Weyerhaeuser's process is a "dry pyrolysis" process (Dehaas et al., 1976; Dehaas, 1979; Hurley, 1980). Several modifications of the dry pyrolysis system were suggested. The simplest of these is appropriate for incremental capacity only. In this process, a portion of the black liquor is pyrolyzed in a fluidized bed. The temperature is kept below the melting point of the inorganics in the liquor. The fuel gas which results is then fired in an auxiliary boiler or in the lime kiln (Figure 2.3) (Whitty and Verrill, 2004). The char would be fed to an existing recovery boiler for reduction of sulfate along with the recovery boiler's char. This type of system can impart more capacity for recovery boilers which have limitations in the heat transfer/flue gas sections.

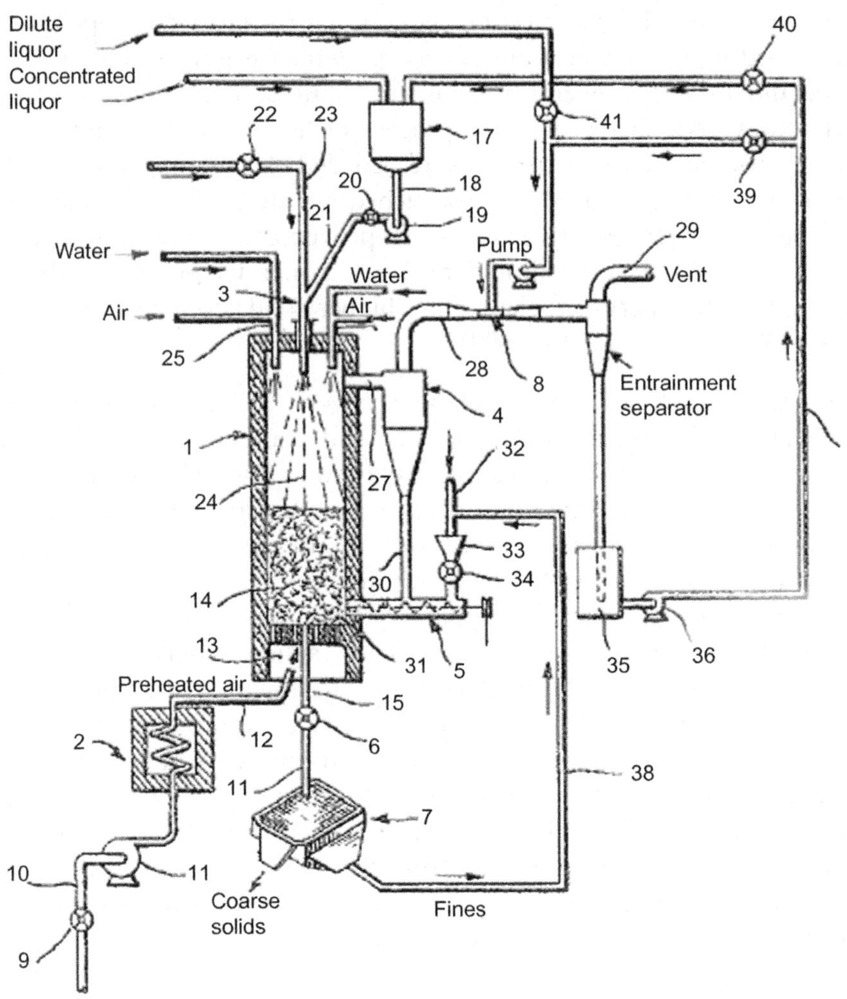

Figure 2.2 The Copeland recovery process. (Copeland and Hanway, 1967).

Other types of the system were also designed. These designs offered an alternative to the recovery boiler which can handle the full load of the black liquor for a mill. The black liquor is fed to a fluidized bed pyrolyzing unit. It forms a char that is passed to another fluidized bed where it is gasified. Fuel gas which is produced in the pyrolyzer is fed to a power boiler. Air is preheated by heat exchange against the inorganic product exiting the gasifier and fed to the gasifier. For black liquors from the kraft process, the gasifier was run substoichiometric and heat for the pyrolysis reactor was supplied by a burner combusting

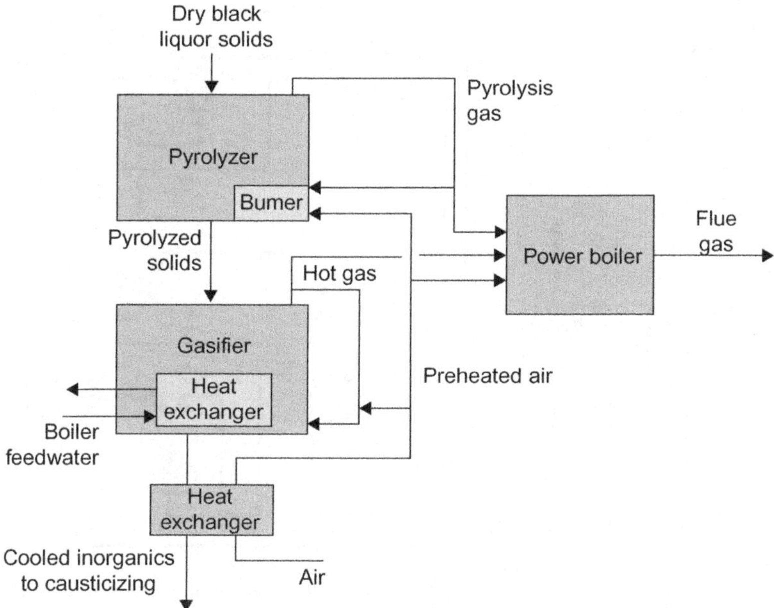

Figure 2.3 Weyerhaeuser's dry pyrolysis process. Reproduced with permission from Whitty and Verrill (2004), adaptation of Dehaas et al. (1976).

a portion of the syngas. In case of nonkraft liquors, the gasifier was meant to operate air-rich and the product from gasifier would be fed to the pyrolysis unit.

2.1.4 The St. Regis Hydropyrolysis Process

The search for this process started in the beginning of 1960s and by 1968, hydropyrolysis was recognized as a viable process for treating black liquor. It was found that heating black liquor under very high pressure produces a slurry which consist of low ash char and alkali salt solution (Timpe, 1973; Timpe and Evers, 1973; Myers and Miller, 1976; Watkins and Timpe, 1980). Figure 2.4 shows the St. Regis hydropyrolysis process (Whitty and Verrill, 2004). Black liquor of low solid content, about ~25%, is pumped to about 2800 psi and heated to a temperature of about 260°C by heat exchange against the product from the reactor. An extra heater raises the liquor temperature to about 332°C. The liquor is then fed to the hydropyrolysis reactor. The residence time is about 5–20 min. The resulting slurry is then cooled and depressurized. Gases are produced during reaction and pressure reduction. These are separated in a flash tank and may be

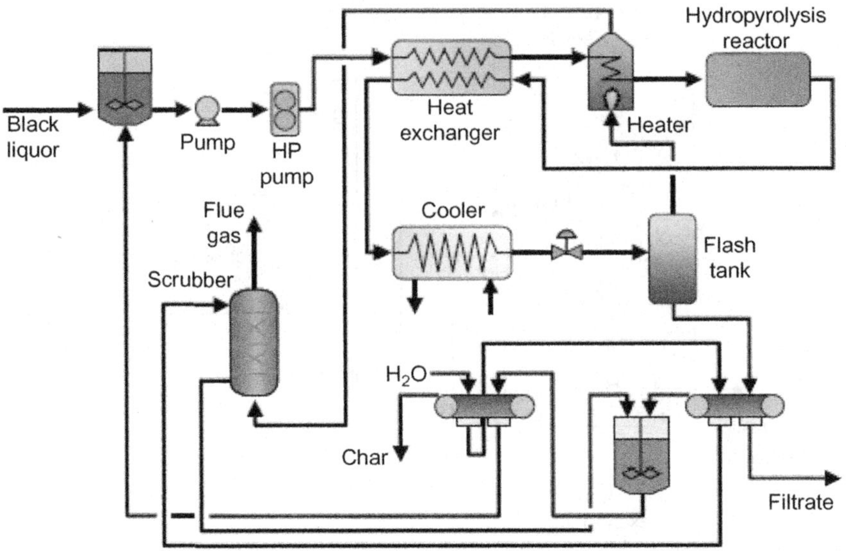

Figure 2.4 The St. Regis hydropyrolysis process. Reproduced with permission from Whitty and Verrill (2004), adaptation of Watkins and Timpe (1980).

used to fire the reactor preheater. The slurry is passed to filters in series where char is separated from the solution and washed to remove sodium and sulfur. The char can be used as a fuel in an auxiliary boiler. It can also be treated to form activated carbon. Early reports of the hydropyrolysis system emphasized the utility of the activated carbon in cleaning process water from the mill. This allowed almost effluent free operation. The filtrate from the slurry is combined with the char washings and forms the green liquor that is supplied to the recausticizing system.

St. Regis carried out laboratory trials between 1968 and 1972, and a pilot plant which processed about 16 tDS per day black liquor was operated in the mid 1970s. Later, engineering for a commercial system with a capacity of about 530 tDS per day black was carried out, with St. Regis' Jacksonville mill. But the plant was never built. The later development work on the hydropyrolysis process was funded by United States, Department of Energy. The development of the hydropyrolysis system was continued till 1980s, but the process was not found to be competitive with the recovery boiler (Grigoray, 2009). Later the research focused on Champion/Rockwell gasifier.

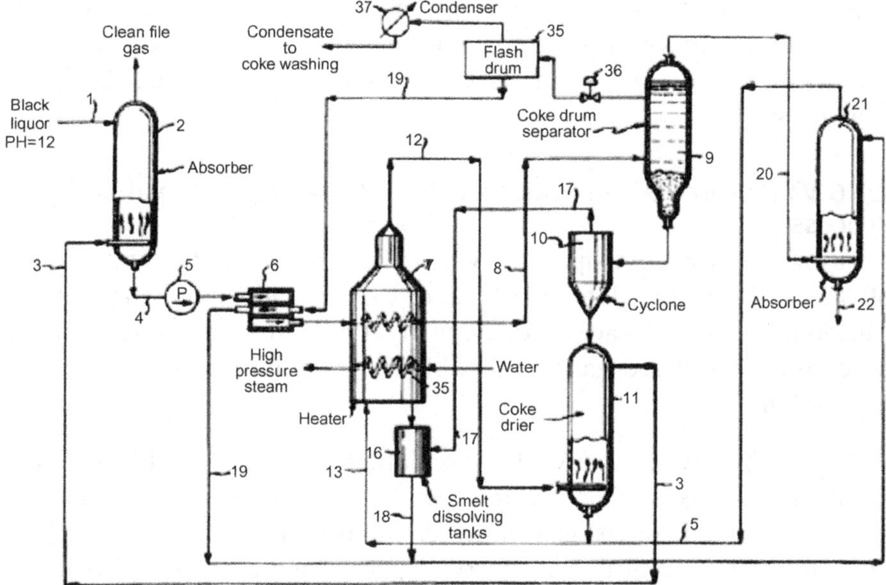

Figure 2.5 The Texaco process. (Hess et al., 1976).

2.1.5 The Texaco Process

This process was quite similar to St. Regis' hydropyrolysis process. Texaco developed this process in the late 1960s and early 1970s. Several patents were filed for this process (Hess et al., 1976, 1978; Hess and Cole, 1971) but not much efforts were made for commercialization of this technology (Whitty and Verrill, 2004).

Figure 2.5 shows the Texaco process. In this process the liquor is treated in liquid phase at high pressure and temperature for 30−60 min to form a solid char, hydrogen sulfide gas and a liquid liquor phase. The pressure and temperature used were 1200 psi and 316°C, respectively. Earlier, the black liquor was taken directly from the blowdown after the digester, pumped to high pressure and then fed to the heated coking reactor. Afterwards a step was added wherein the pH of the liquor was reduced by treating it with sulfur dioxide produced in the boiler. This allowed coking to proceed considerably faster and at lower temperature. Upon leaving the coking reactor, gas containing much of the sulfur in the form of hydrogen sulfide, was flashed from the effluent and typically fed to a power boiler. The liquid slurry was washed to separate the liquor containing the pulping chemicals from the coke. The liquid effluent was treated in the same way as green liquor from a conventional recovery process. The coke was washed and dried, and then, it was used as fuel in a boiler to increase steam.

2.1.6 VTT's Circulating Fluidized Bed Black Liquor Gasification Process

This process was developed by VTT (The Technical Research Center of Finland) in the late 1980s and early 1990s (Saviharju, 1993; Mckeough, 1993; Grace and Timmer, 1995). The objective was to develop a pressurized gasifier suitable for a combined cycle cogeneration system.

Figure 2.6 shows the VTT's circulating fluidized bed black liquor gasification process. This is a low temperature process and is basically a pressurized, air- or oxygen-blown circulating fluidized bed, operating at about 650°C (Whitty and Verrill, 2004; Grigoray, 2009). Preheated air is passed through a gate at the bottom of the bed, and reacts with the fuel in the bed to provide the heat required for gasification. The bed material was intended to be the salt residue from liquor

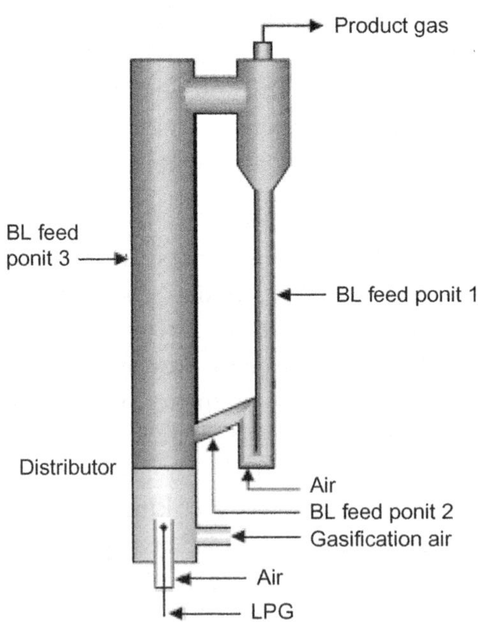

Figure 2.6 VTT's circulating fluidized bed BLG process. Reproduced with permission from Whitty and Verrill (2004), adaptation of Saviharju (1993).

conversion, but the bed material was changed to Al_2O_3 in VTT's pilot study due to problems with bed melting. Laboratory studies were conducted to identify partitioning of species between the gas and solid phases and reaction kinetics. Concurrently, a pressurized, 1.5 tDS per day development unit was built at a mill in Äänekoski. This facility had three entirely different pilot gasifiers: VTT's CFB gasifier, a high temperature, pressurized black liquor gasifier being developed by Ahlstrom and a high-temperature, atmospheric pressure entrained flow sulfate soap gasifier developed by VTT. All the three reactors shared common subsystems for feedstock preparation and pumping, product gas handling, and air, nitrogen, steam, water and LPG supply. This resulted in reduced capital and engineering burden for the partners, but the disadvantage was that only one reactor could be operated at a time. Because of the technical problems with the gasifiers and the economic condition in Finland, the whole testing program was ended at the end of 1992.

2.1.7 Babcock and Wilcox's Gasification Process

Babcock and Wilcox studied fluidized bed black liquor gasification. The project was sponsored by US DOE (US DOE, 1997). The bubbling fluidized bed reactor is operated at a temperature of about 650°C, below the melting point of the inorganic residue from the liquor. Initially atmospheric pressure gasification was used. In this process, black liquor is sprayed from the top of the reactor, above the freeboard, and falls onto the bed. The liquor dries and partly pyrolyzes as it falls through the freeboard. The bed is fluidized using steam, air and hot gases which result from the combustion of syngas produced in the gasifier. Injector is used to introduce the gas into the bottom of the reactor. The reactor is of special design. The researchers hold patents on the design. The design is a gas burner which is surrounded concentrically by a steam injector, all encased in a casing or bubble cap. The fuel, air, and steam flows are adjusted. This allows control of the temperature of the incoming gas, which controls the temperature of the reactor. The composition of the reacting gas can be controlled. An air-blown bench-scale gasifier was built to test important aspects of the system. The studies were successful (Whitty and Verrill, 2004). There were some running problems relating to the feed system and defluidization by char agglomerates. These were mainly due more to the small scale and particular design of the system. It was concluded that LTBLG shows potential. However, further work at larger scale is necessary to demonstrate the technology. The researchers developed a basic design for a 1000 tDS/h pilot plant but the pilot plant was not built (Dickinson et al., 1998; Kitto, 1997; Mcilroy et al., 1997).

2.1.8 NSP Process (Ny Sodahus Process)

Extensive work has been done in Sweden on a cyclone gasifier (Figure 2.7) for black liquor that is mounted to a Tomlinson recovery boiler as a prefurnace (Whitty and Verrill, 2004; Björkman, 1976; Holme, 1975, 1976; Magnusson and Warnqvist, 1980; Empie, 1991; Whitty, 2005). This configuration provides for significantly better control of the flows of liquor, gases, and smelt, along with improved mixing and heat and mass transfer characteristics. Energy and reduction efficiencies, as well as unit productivity, are expected to be improved with the NSP. Black liquor and a limited amount of air are fed to the cyclone, where the basic steps of drying, pyrolysis, gasification, and combustion are performed. A reducing atmosphere exists at the outlet or point of attachment to the recovery boiler, where fully reduced smelt leaves the cyclone and flows to the dissolving tank. The flue gases flow through the recovery boiler and are fully combusted by secondary air. Heat recovery is accomplished in the usual way in the Tomlinson boiler. When mill scale tests of a full-size cyclone on a standby recovery boiler were run, they identified a critical problem: supplemental fuel was needed to maintain the high temperatures required for the endothermic physical and chemical processes which

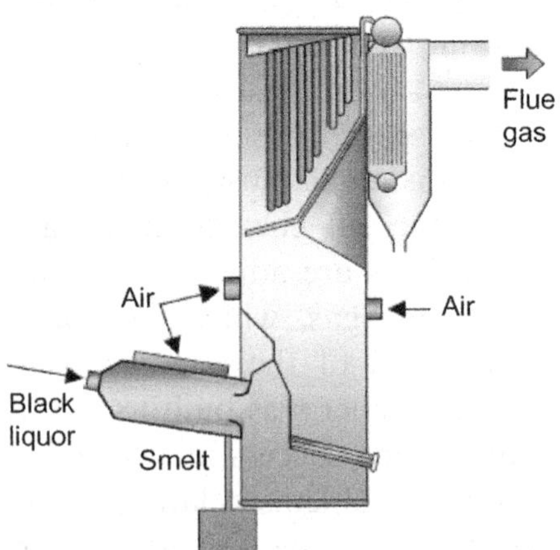

Figure 2.7 NSP cyclone gasifier. Reproduced with permission from Whitty and Verrill (2004), adaptation of Magnusson and Warnqvist (1980).

occur in the low residence time cyclone. Continued development of the NSP concept awaited a financial commitment by a potential industrial user. One pilot unit was built and operated for a few years in the beginning of the 1980s. Technical difficulties and lack of funding discontinued the project (Empie, 1991).

2.1.9 DARS Process

The Direct Alkali Recovery System (DARS) is one of the first alternative recovery concepts.

It is based on the addition of an *in situ* causticizing agent to spent pulping liquor prior to combustion (Grigoray, 2009; Tadashi et al., 1976; Covey and Ostergren 1985; Kulkarni et al., 1987; Rao and Kumar, 1987; Venkoba Rao, 1987; Bajpai, 2008). This process shows promise of providing a simpler black liquor recovery technology that is suitable for small-scale pulp mills. In the basic process, patented by the Toyo Pulp Company in 1976, iron oxide combines with sodium carbonate to form sodium ferrite, which expels carbon dioxide to the flue gases. When the sodium ferrite is dissolved in water, the salt decomposes to form sodium hydroxide plus insoluble iron oxide which can be removed from the caustic liquor and recycled to the fired black liquor (Tadashi et al., 1976):

$$Na_2CO_3 + Fe_2O_3 = Na_2Fe_2O_4 + CO_2$$

$$Na_2Fe_2O_4 + H_2O = 2NaOH + Fe_2O_3$$

Coarse particles (1−3 mm in diameter) are maintained to allow more efficient de-watering of the ferric oxide after leaching. The fluid bed is a "bubbling bed" type combustor to minimize particle size reduction, which would occur with a circulating bed (Grace, 1987). The obvious advantage of this process is the elimination of the lime-based causticizing process, which requires fossil fuel for driving the lime mud calcination reaction. The DARS process is limited in application to non-sulfur pulping because iron reacts with the reduced sulfur to form undesirable compounds that cannot be recovered in useable forms for reuse in pulping. The commercial realization of the DARS process combined this process chemistry concept with a Copland-like fluidized bed combustor and associated solids-handling equipment. The concept was licensed from Toyo and successfully demonstrated by what is now Australian Paper in the early 1980s using an integrated pilot plant including a batch digester, spent liquor evaporator, and a 76 cm diameter fluidized bed combustor.

In 1986, a full DARS plant was constructed by Associated Pulp & Paper Mills (APPM) to replace rotary incinerators at its Burnie, Tasmania soda process pulp mill. The fluidized bed combustion zone operated at approximately 1000°C. Because there was no sulfur in the process, it was not necessary to operate in a reducing mode. The sodium ferrite bed solids were continuously removed from the combustor and leached by water in a countercurrent contactor. The equilibrium concentration of sodium hydroxide from sodium ferrite hydrolysis could reach levels two to three times as high as with conventional lime-based causticization. Solids were separated from the caustic liquor, mixed with make-up iron oxide, and returned to the fluidized bed. An expected challenge to operating this process was managing the fine dust generated from attrition of the porous leached bed solids. Accordingly, a baghouse filter was used to remove the dust from the flue gas, and this material was agglomerated by mixing it with a small amount of black liquor prior to recycling it to the fluidized bed combustor.

Difficulties were faced during the first five years of operation, especially in maintaining fluidization and managing the large quantities of sodium ferrite dust. The system was able to operate for an extended period only after three major equipment rebuilds. By 1995, the Burnie mill had been acquired by Australian Paper, and the DARS plant was processing about 150 tons of dry solids per day and supporting two-thirds of the mill production. By the late 1990s, the process was running well enough that a second DARS plant and expansion of the mill was under evaluation, but a decision was reached that, even with a capacity increase, the pulp mill was too small to be economically viable and so the entire operation was shut down.

Several companies have evaluated DARS for use in nonwood pulping operations, and at least two additional pilot plants have been built, but a second commercial unit has not yet been commissioned. The DARS process shows promise for providing a simpler recovery technology suitable for smaller scale mills. The economic analysis indicates that the system offers a substantial reduction in both the capital investment and operating costs compared with conventional recovery systems. One drawback of the process is that it is not applicable to kraft recovery because the iron oxide is reduced under the conditions needed to form sulfide (Rao and Kumar, 1987).

2.1.10 BLG with Direct Causticization

Direct causticization is one of several possible processes that can be used to recover the pulping chemicals in black liquor. It has been shown that direct causticization has several advantages to conventional lime-based recovery processes, including higher production of electricity, and a smaller compounds of nonactive chemicals in the white liquor which is recycled back to the digester or pulping stage. However, the chemical composition of the white liquor produced via direct causticization development differs from that of the lime-based processes and could impact the pulping yield and pulp quality (Naqvi et al., 2010).

ABB developed a new BLG technology with direct causticization for energy optimization during the 1990s (Dahlquist and Jacobs, 1994). Figure 2.8 shows ABB circulating fluidized bed BLG process. A circulating fluidized bed of titanium dioxide is injected with black liquor under pressurized conditions (5 bar) and temperature range 650–720°C. The titanium dioxide reacts with sodium carbonate to form sodium titanate and carbonate is converted to carbon dioxide that results in direct causticization. First, sodium sulfate is reduced to

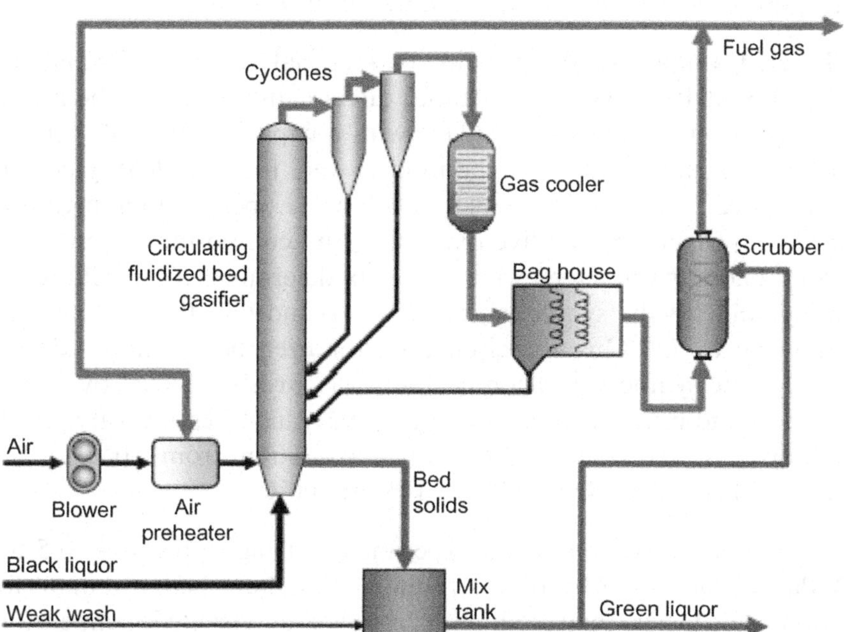

Figure 2.8 ABB circulating fluidized bed gasification process. Reproduced with permission from Whitty and Verrill (2004), adaptation of Dahlquist and Jacobs (1992).

sodium sulfide, and then sodium sulfide is stripped off as hydrogen sulfide. The hydrogen sulfide is absorbed from the raw gas in a selective absorber, using part of the white liquor for absorption. ABB continued development of their black liquor gasifier until 1997, when the program was abandoned due to shifting development priorities and lack of a clear market for the gasifier in a reasonable time frame.

Several researchers have studied BLG process with direct causticization (Dahlquist and Jacobs, 1994; Dahlquist, 2003; Zou et al., 1992; Zeng and van Heiningen, 1996; Nohlgren, 2004; Nohlgren and Sinquefield, 2004). Dahlquist and Jacobs (1994) studied BLG process with direct causticization using equilibrium calculations. The experiments were conducted in a 1 ton BLS per day pilot plant in Vasteras. The formation of CO and H_2 in the synthesis gas was lower than expected but significantly higher for CH_4. Equilibriums were calculated at different operating conditions in order to get concentration of residual carbon, CO, CO_2, H_2, CH_4, H_2S, and H_2O. With only 35% relative oxidation for organics in the black liquor, a synthesis gas rich in H_2, CO, and CH_4 was produced. It was estimated that if the heat in the synthesis gas was used to generate steam for a steam turbine, up to 38% electrical efficiency can be obtained from the black liquor heating value, more than normal recovery boiler cycle (9–14%).

Later, Dahlquist (2003) presented a factorial design model and controllability of BLG with and without direct causticization. The models were based on tests in a pilot plant with a capacity of 1 tDS per day. The experimental results and the models were used to identify correlations between important variables. These were operating temperatures, production capacities, relative oxidation, and composition of the black liquor. Those parameters were used to determine composition and heating value of the synthesis gas. It was found that the process could operate up to 720°C in the circulating fluidized bed without addition of TiO_2. The synthesis gas composition was largely affected by relative oxidation and must be kept low to achieve a high heating value of the synthesis gas. An increase in relative oxidation from 30% to 50% decreased H_2 content from 11% to 10% at 700°C.

Nohlgren (2004) conducted experiments at high pressures (0.5 and 1 MPa) at 900–1000°C. It was found that sodium penta-titanate formation was faster than sodium carbonate at elevated temperature. There was no clear effect of pressure on the rate of reaction but carbon dioxide formation slowed down the reaction rate. Nohlgren and Sinquefield (2004) studied the influence of molar ratio of $2TiO_2/Na_2O$ on causticization and also carried out equilibrium calculations. For a complete causticization, a molar ratio of 0.5 was required to capture sodium in the condensed phase as sodium titanates that could be further used to produce sodium hydroxide. Zou et al. (1992) and Zeng and van Heiningen (1996) studied BLG with direct causticization using medium temperature fluidized bed operation at 800–850°C and atmospheric pressure.

Most of the studies showed that the direct causticization has several advantages over the conventional black liquor recovery system including lower capital cost for fewer process steps, e.g., no requirement for the lime kiln (Warnqvist et al., 2001; Sinquefield, 2005). A higher concentration of white liquor with different sulfur contents can be achieved. In terms of energy efficiency, the direct causticization process is better than conventional recovery system; only 25% of the energy demand in conventional recovery is required for direct causticization (Richards et al., 2002).

2.1.11 Manufacturing and Technology Conversion International Fluidized Bed Gasification

This process was developed by MTCI and commercialized by TRI. This is an LTBLG process which is carried out at a temperature below the melting temperature of inorganic chemicals present in black liquor. In this gasification technology, TRI applies the indirect heating of black liquor in a steam fluidized bed at a temperature of approximately 600°C and at atmospheric pressure. As gasification equipment, TRI utilizes a steam reformer (Figure 2.9), which has two main output products: synthesis gas and alkali salts as solids. Indirect gasification creates conditions of organic substances converting into product gas in the absence of air or oxygen. Since steam plays a role of gasification and the fluidizing agent and transfers its energy to black liquor, the need of direct combustion of the feedstock in the presence of air or oxygen disappears (KBLG, 2006; Durai-Swami et al., 1991; Mansour et al., 1992, 1993, 1997; Rockvam, 2001; Whitty and Verrill, 2004; Whitty, 2005; Grigoray, 2009). MTCI projects are running in mills with a Na_2CO_3 semi-chemical cooking process. It operates at 600–620°C. This technology has some significant advantages (Table 2.3).

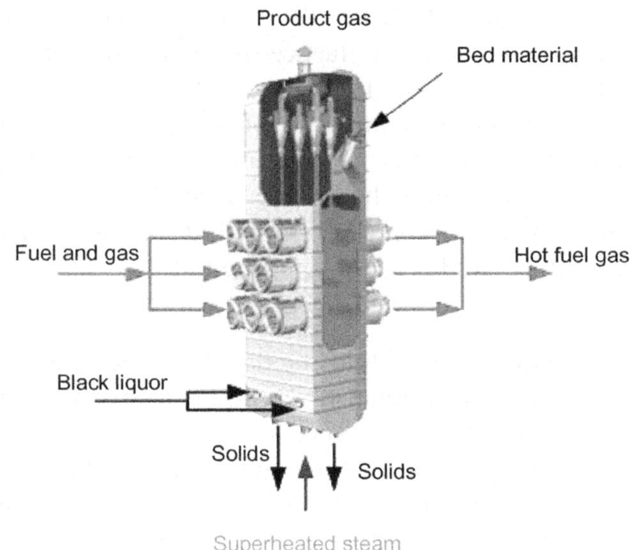

Figure 2.9 TRI steam reformer based on Grigoray (2009) and TRI (2013).

Table 2.3 Advantages of MTCI/TRI Technology
Efficient heat transfer with uniform heat flux
High combustion efficiency
Low NO_x emissions
Absence of moving parts
Pressure boost helps exhaust gases to flow through a super heater and heat recovery steam generator (HRSG)
Source: *Based on Naqvi et al. (2010).*

There are two MTCI projects running. The first one is running at Georgia Pacific Corporation's Big Island mill in Virginia. Georgia Pacific Mill in Big Island, having a capacity of 1000 tons per day of linerboard and 600 tons per day of corrugating medium, uses a sodium carbonate process. The smelters used, which were initially installed, had the function of chemical recovery but energy recovery was not carried out. In 2001, the indirect gasification process was used instead of the smelters. Initially gas produced was cleaned and then burnt in pulse combustors. The capacity of the steam reformer was 200 tons per day of black liquor dry solids. The system included two reformers and four pulse combustors. The process was shut down in 2007. During work the following weaknesses were identified (Middleton, 2006; Newport et al., 2004; Vakkilainen et al., 2008; DeCarrera, 2006):

- Tar formation leading to plugging problems
- Incomplete conversion of carbon
- Carburization problems.

The second project is running at Norampac Trenton mill, Ontario, Canada. This mill makes 500 tons per day of corrugated board and utilizes sodium carbonate pulping methods. Before the implementation of the gasification process at the mill, there was no recovery system and black liquor produced was bought by other countries, where it was applied as a road spray for dust suppression. In September 2003, a black liquor steam reformer was taken into use. The productivity of the unit is 115 tons of BLS per day. Unlike in Big Island, Norampac uses only one steam reformer and an auxiliary boiler to burn off produced syngas. The first demonstration was completed in December 2003 The main issue during the gasification process was plugging problems. After numerous tests by TRI at both the Big Island and Ontario mills, the steam reformer was relaunched again in April 2004. The results of the gasification process are a 99% recovery of sodium supplying the mill by process steam and with 100% environmental technology. Currently Norampac's steam reformer prolongs its work (KBLG, 2006). In spite of the headway done in the field of BLG, TRI has decided to return to the gasification of traditional biofuels and use its gasifier for these purposes because of the technical problems appearing during BLG process (Whitty, 2009).

This process is ideal for use in a forest products biorefinery as it is uniquely configured for high-performance integration with pulp and paper facilities and is capable of handling a wide variety of cellulosic feedstocks, including woodchips, forest residuals, agricultural wastes and energy crops, as well as mill byproducts (spent liquor) (Connor, 2007). Compared to other biomass gasification technologies that are based on partial oxidation, TRI's steam reformer converts biomass to syngas more efficiently, producing more syngas per ton of biomass with a higher Btu content. This medium-Btu syngas can be used as a substitute for natural gas and fuel oil, and as a feedstock for the production of value-added products such as biodiesel, ethanol, methanol, acetic acid, and other biochemicals. TRI's technology can be integrated with a wide variety of catalytic and fermentation technologies to convert the syngas to high value bio-based fuels and chemicals. For example, syngas generated by TRI's technology can be conditioned and sent to a commercially proven gas-to-liquids (GTL) facility (i.e., FT or other catalytic technologies) inside the biorefinery. The GTL process produces a range of products—naphtha, gasoline, diesel/kerosene, wax, methanol, DME, etc.—that are stabilized for storage and transported offsite to a downstream refinery for conversion to marketable products. The unreacted syngas and light noncondensable gases (tail gas) are utilized in the process to replace fossil fuels. Additionally, the GTL conversion, which is exothermic, provides another source of process heat that is recovered and used. A fully integrated forest products biorefinery utilizing TRI's technology will achieve thermal efficiencies from 70% to 80% depending upon process configuration and biomass feedstock.

A major drawback with MTCI process is low carbon conversion due to low temperature, but still the thermal efficiency is well above 70% as compared to 65% or less for conventional recovery boilers (Suresh, 2002).

The indirect gasification of black liquor via the steam reformer is called liquor steam reforming because this process is based on the reaction of steam with organic carbon (as shown in the below equation) instead of partial oxidation liquor as is usual in the gasification process (Grigoray, 2009).

The steam reformation is an endothermic reaction:

$$H_2O + C + Heat = H_2 + CO$$

The main products of liquor treatment by superheated steam are hydrogen and carbon monoxide. Then steam interacts with carbon monoxide and more hydrogen and carbon dioxide are obtained according to the following equation:

$$CO + H_2O = H_2 + CO_2$$

During the drying and heating of black liquor, a significant amount of hydrogen, carbon monoxide, carbon dioxide, and methane is formed via the release of volatile components (DeCarrera, 2002).

Over 90% of the sulfur compounds in black liquor are converted into sulfide gas under the influence of the superheated steam

(KBLG, 2006). The typical composition and heating value of syngas obtained by black liquor steam reforming after its separation from hydrogen sulfide is presented in Table 2.4.

Carbon dioxide produced during liquor steam reforming combines with potassium and sodium hydroxide and form carbonates (sodium carbonate and potassium carbonate) in the form of solids. The results of the conversion of black liquor in the steam reformer can be presented as follows:

- The production of hydrogen-rich and medium Btu synthesis gas
- The almost complete segregation of sulfur compounds from alkali.

The indirect gasification process is based on converting the organic components of black liquor into syngas without direct combustion and executed by means of the steam reformer used as gasifier. The bed material, which is sodium carbonate, is introduced in the steam reformer first. The size of sodium carbonate particles varies within the limits of 100–600 microns. The bed material serves as a catalyst promoting a more intensive interaction of the gasifier agent which is steam with the feedstock (liquor) by increasing the reaction surface area and thus raising the reformer capacity. After filling the reformer vessel with sodium carbonate, superheated steam is fed into the vessel, preliminary going through the super heater. The steam has two functions:

1. Fluidizing bed material
2. Heat transfer from the source of heat to the feedstock.

Pulsed combustion heat exchangers are the source of indirect heating. A mixture of fuel and air, the flow rate of which is adjusted by aerovalves, enters into the combustion chamber and is ignited by a

Table 2.4 Composition and Heating Value of Cleaned Syngas Obtained by Black Liquor Steam Reforming	
Carbon monoxide (%)	23.7
Carbon dioxide (%)	10.5
Hydrogen (%)	61.9
Methane (%)	3.5
Heating value (MJ/kg)	20.95
Source: Based on Grigoray (2009).	

pilot flame. The incineration of this mixture leads to its expansion and fuel gases are pushed into resonance tubes and then leave the heater. When fuel gases leave the combustion zone, a vacuum is formed bringing more fuel and air into the chamber and also the phenomenon of the reverse motion of gases remaining in the tubes. A fresh mixture is ignited by returning hot gases and the process repeats itself. The frequency of the process (pulsations) is 60 times a second. The heat transfer efficiency of the heater is provided by perpendicular arrangement of its pipes to the streams of bed materials and steam, and also by a permanent change of the direction of flue gases (TRI, 2013). Black liquor is fed into the vessel after the bed reaches operation temperature, which is lower than the slugging temperature of liquor components. As soon as the feedstock enters the reformer, water contained in the liquor evaporates and then volatile components are liberated. Inorganic salts in the form of sodium carbonate and calcium (present in a small amount) are discharged from the reformer as dry solids. Cyclones installed at the top of the reformer are intended to separate solids taken away by produced gases. Figure 2.9 demonstrates the work of the reformer. In addition to carbonates, the outgoing solids contain nonprocess elements (calcium and silicon) and unburnt carbon. These are removed by countercurrent washing and filtering after solids dissolution. The cleaned alkali solution is used in pulp mills for the preparation of white liquor and undesired elements are discharged as dregs. Figure 2.10 shows cleaning system of syngas and green liquor generation (DeCarrera, 2002). The syngas leaving the gasifier passes through a cleaning and sulfur recovery system. The gas is first quenched and impregnated with water in a heat recovery steam generator (HRSG) (Figure 2.10). Then undesired particles are moved off by scrubbing and the gas goes to cooling. Purified solids and chilled gas is sent to the multistage countercurrent scrubber where it loses sulfur hydroxide by absorption in a mixture of sodium carbonate and caustic solution producing green liquor (DeCarrera, 2002). The indirect gasifier is self energy efficient since the syngas and steam produced during the process are the main inputs needed for the reformer work. Product gas and fuel gases are sent to the reformer boiler for steam generation. The main issue of the steam reformer working properly is keeping the required temperature, which in turn decreases the probability of tar formation that causes clogging and plugging and avoids the agglomeration of bed material (KBLG, 2006).

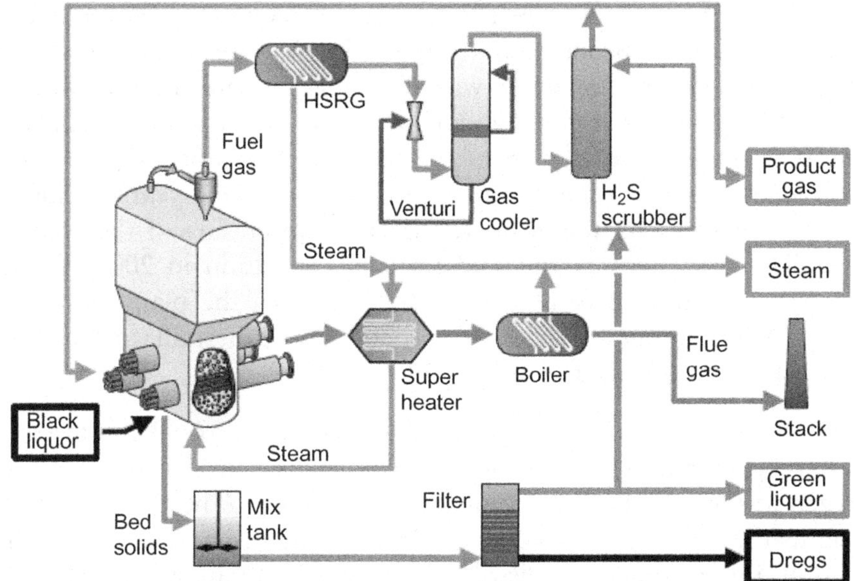

Figure 2.10 MTCI system configuration for kraft liquor. Reproduced with permission from Whitty and Verrill (2004), adaptation of MTCI (1997).

2.1.12 Chemrec Gasification

This technology is currently the most commercially advanced BLG technology and is based on entrained flow gasification of the black liquor at temperatures above the melting point of the inorganic chemicals. Chemrec is working on both an atmospheric version and a pressurized version of a high temperature downflow entrained flow reactor. The atmospheric versions are mainly considered as a booster to give additional black liquor processing capacity. The pressurized version is more advanced and would replace a recovery boiler or function as a booster (Brown and Landälv, 2001; Kignell, 1989; Stigsson, 1998; Whitty and Nilsson, 2001; Whitty and Verrill, 2004).

2.1.12.1 Atmospheric System

This technology allows for capacity additions in a mill where the recovery boiler is already running at maximum capacity. It is designed to handle the extra black liquor formed as a result of the increase in capacity. The booster technology is simpler; it is not pressurized, uses air as gasification agent, and produces a fuel gas that is used for steam production, which means that it can be considered as a commercial technology. As early as 1995, when the US company Weyerhaeuser wanted to increase production at its mill in New Bern and invested in a Chemrec booster, the facility was able to supply green liquor with an acceptable quality (the green liquor may be of lower quality because it only represented a small part of the total green liquor) (Figure 2.11). The facility has had some technical problems over the years, including material problems in the gasifier, which led to closing of the plant in 2000. The plant was rebuilt and resumed operation in 2003 (Whitty and Nilsson, 2001; Brown et al., 2004). In 2008, the plant was again closed, this time due to decreased production at the mill, and has not resumed operation since.

In this system, black liquor is fed as droplets through a burner at the top of the reactor. The droplets are partially combusted with air or oxygen at 950–1000°C and atmospheric pressure. The heat generated sustains the gasification reactions. The salt smelt is separated from the gas, falls into a sump, and dissolves to form green liquor. The produced gas passes a cooling and scrubbing system to condense water vapor and remove H_2S. The gas has low heating value ($\sim$2.8 MJ/Nm^3) and is suitable for firing in an auxiliary boiler. It consists of 15–17% CO_2, 10–15% H_2, 8–12% CO, 0.2–1% CH_4, and 55–65% N_2 (Lindblom, 2003). The thermal efficiency is quite low.

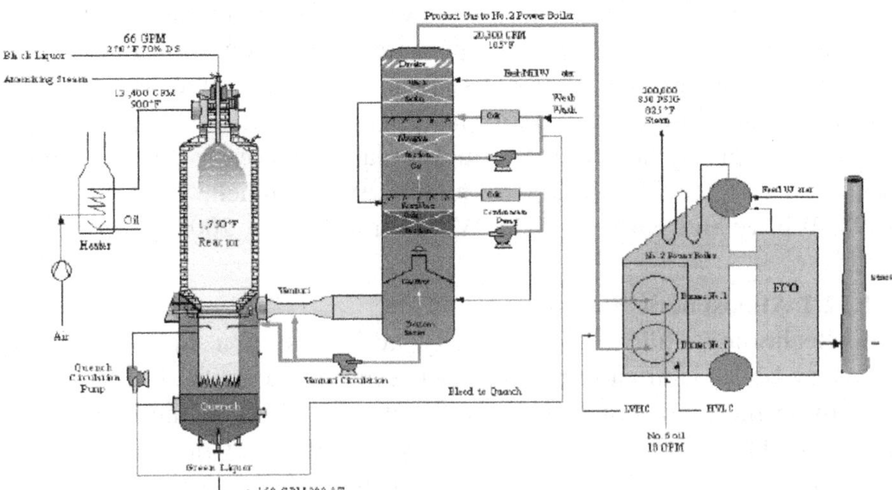

Figure 2.11 Chemrec booster gasification system, Weyerhauser Mill, New Bern, NC. Reproduced with permission from Chemrec.

2.1.12.2 Pressurized System

The pressurized CHEMREC BLG system is a new technology for energy and chemicals recovery in pulping processes with the aim of offering pulp mills significant cash flow additions through increased utilization of the energy content in its renewable feedstocks for the production of green electricity, automotive fuels, or hydrogen. In addition, the CHEMREC system offers the option of utilizing advanced new cooking processes resulting in increased pulp yield and giving a fundamentally different opportunity to manage the environmental impact of black liquor conversion. The capital outlay for the system is larger than for the corresponding recovery boiler, but the extra investment is rapidly repaid through the increased cash flow generated.

The core of the pressurized CHEMREC system is the gasifier unit, an entrained flow reactor, where concentrated black liquor is gasified at approximately 32 bar pressure and at a temperature of about 1000°C. Oxygen is used as the oxidant. In the CHEMREC Demonstration Plant 1 (DP1), plant operations start with a refractory lined reactor with the option of later switching to a reactor with a cooling screen making up the containment. Black liquor reacts in the reaction zone to form smelt droplets consisting of sodium and sulfur compounds and a combustible gas mainly consisting of CO, H_2, and CO_2. Part of the sulfur in the black liquor also ends up as H_2S in the gas. The smelt droplets and the combustible gas are fed to a quench vessel where they are cooled when brought into direct contact with condensate, which is recycled from the downstream gas cooler. The smelt droplets are separated from the gas and dissolved in the quench liquid to form a green liquor solution. The green liquor is after heat exchanging with weak wash and final cooling fed to battery limit. The gas leaving the primary quench is further cooled to saturation in a second quench device. The saturated gas is simultaneously scrubbed from particles and cooled in a CCC. The DP1 plant utilizes cooling water as cooling medium in the CCC. In mill applications, the CCC will instead produce medium- and low-pressure steam. The cooled gas is cleaned from sulfur in multistage short time contactors utilizing white liquor as absorbing liquid after which the purified and cooled syngas is burned in a flare. The CHEMREC reactor is designed to achieve high carbon conversion and sulfur reduction, exceeding what is normally obtained in a recovery boiler. The quantity of unburned carbon and sulfate in the green liquor is consequently low. The CHEMREC DP1 plant is operated from its own control room through a state-of-the-art computer-based steering and control system. Media such as black liquor, white liquor, steam, water and electricity are supplied from the adjacent Kappa Kraftliner mill and green liquor is delivered back to the mill. The DP1 plant operation is managed by CHEMREC and during longer periods of operation supported by staff from the Kappa Kraftliner and SCA Packaging Munksund mills (Grigoray, 2009). Figure 2.12 shows the Chemrec DP1 plant.

During the development of the Chemrec technology, several different plants have been operated during the last decades (in Hofors, Frövi, Skoghall, New Bern, and Piteå) (Figures 2.13–2.15). The development plant in Piteå was taken into operation in 2005 (Bergek, 2002; ETC, 2011). The development has gone from nonpressurized air-blown gasification to pressurized oxygen-blown gasification. Since both firing in a gas turbine for electricity generation and motor fuel synthesis occur at high pressure, it is advantageous if the synthesis gas is already pressurized. Furthermore, pressurized gasification offers the advantages of smaller equipment and the ability to produce low- and medium-pressure steam from gas cooling, which can be used in the pulping process (Stevens, 2001; Whitty and Nilsson, 2001). Air-blown gasification can be used if electricity generation is envisaged. However, the gas from air-blown gasification has a low heating value because

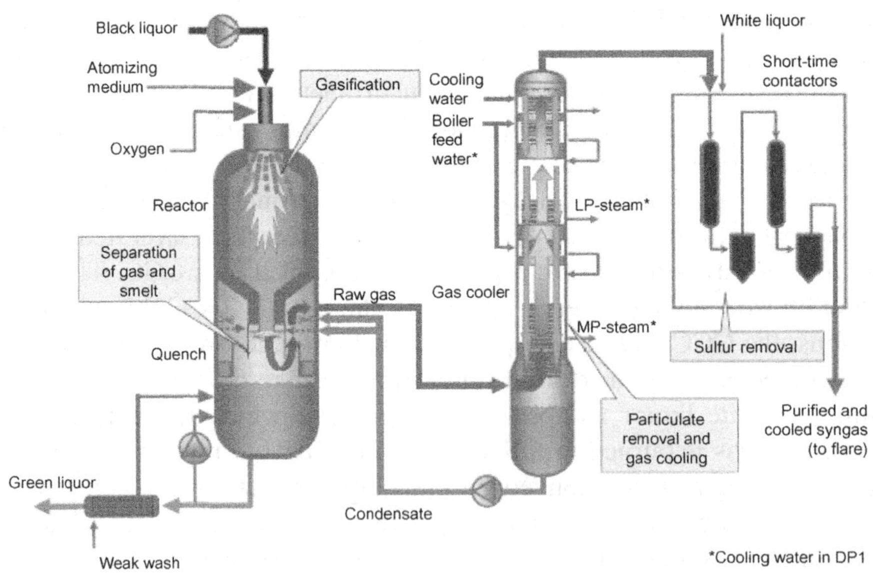

Figure 2.12 The CHEMREC DP1 plant. Reproduced with permission from Chemrec.

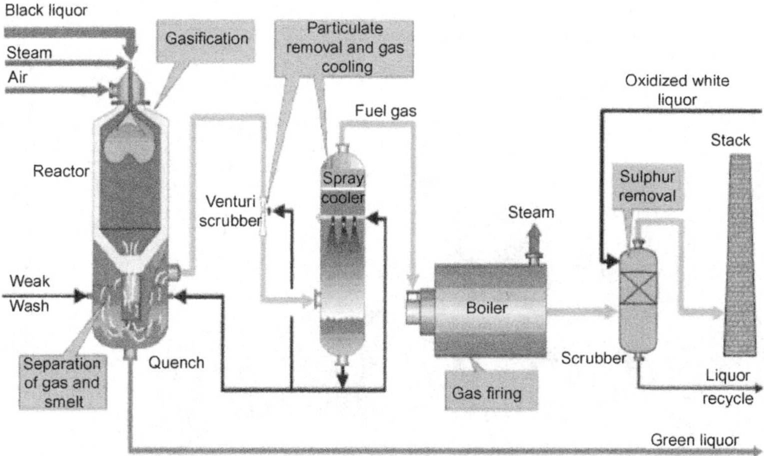

Figure 2.13 Chemrec air-blown gasification system. Reproduced with permission from Chemrec.

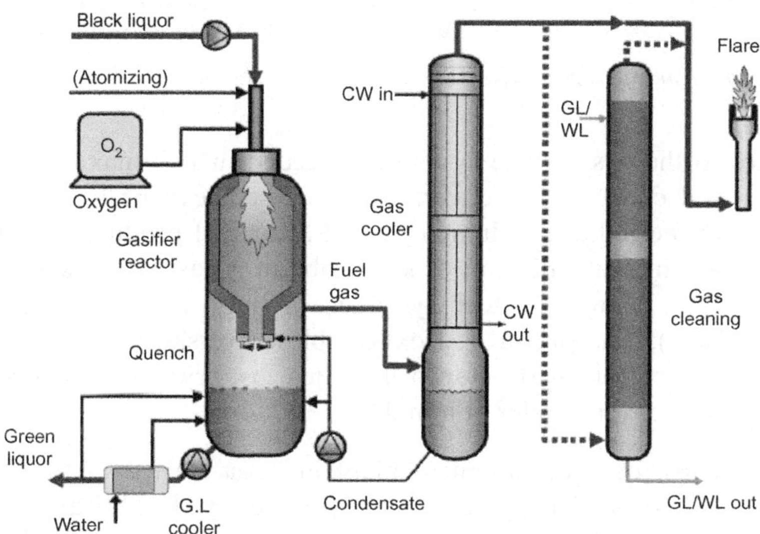

Figure 2.14 Chemrec gasification system, Stora Enso Mill, Skoghall, Sweden. Reproduced with permission from Chemrec.

it contains significant amounts of inert gas (mainly nitrogen), which requires modifications of the gas turbine combustion unit. Furthermore, it is very difficult to start the gas turbine system and maintain stable operation. These problems do not occur if oxygen is used instead of air for the gasification. In this case, the gas obtained has a higher heating value as it contains less inerts and no structural

Figure 2.15 The DME plant in Piteå. Reproduced with permission from Chemrec.

changes to the gas turbine system are needed, which makes the plant simpler and cheaper. Moreover, the equipment can be made smaller because of the reduced volume of gas. Motor fuel production requires oxygen-blown gasification in order to obtain a gas with a sufficiently high heating value for fuel synthesis (Whitty and Nilsson, 2001; Marbe, 2005). The pressurized oxygen-blown version of the Chemrec technology, currently developed in Piteå, is primarily intended to replace the recovery boiler at a mill.

The plans for a development plant in Piteå started in the second half of the 1990s. During the 1990s, BLG received political attention as a possible technology for increased electricity production in the pulp and paper industry which could contribute to increased renewable electricity production in Sweden. This finally led to a Swedish government body deciding to support construction of a BLGCC development plant (Bergek, 2002). For a number of reasons, such as changed ownership structure in both Chemrec and the mill in Piteå, the plans were delayed and restricted to a plant for gasification and gas cooling (Bergek, 2002). Some of the reasons for which Piteå was selected as location for the development plant were significant interest from the mill owner (a kraftliner mill now owned by Smurfit Kappa) for the

BLG technology and availability of a suitable space for the plant at the adjacent Energy Technology Research Centre ETC Piteå (Energitekniskt centrum i Piteå). The plant, which started operation in September 2005, has a capacity of 3 MW black liquor (20 ton dry solids per day), corresponding to approximately 1% of the black liquor produced at the mill. In the development plant in Piteå, the standard gasifier operating conditions are 29 bar, 1000°C, and the black liquor from the mill has a normal dry solid content of 73% (Landälv et al., 2010). In 2004, a research program about BLG was started, the Black Liquor Gasification Program, which was a continuation of a program that started in 2001. During the first phase of the program (BLG I), 2004–2006, the task was to build, commission, and test the development plant in Piteå, and perform fundamental research on a number of issues connected to the technology. During the second phase (BLG II), 2007–2010, the overall goal was to remove scientific obstacles to commercialization of BLG, to understand the process and to place it in its context within the pulp mill. Research has been conducted at several universities and research institutes in Sweden with financial support from agencies and foundations, as well as industry (ETC, 2011). The program worked on, for example, challenges regarding materials for the gasifier, which has resulted in materials that can handle several years of operation. The problems with increased concentration of nonprocess elements are not as great as was expected, and there are different possible solutions to the problem. The results from the plant indicate an increase of the need for CaO, and correspondingly also the load of the lime kiln, by approximately 33% (ETC, 2011). This is lower than predicted value − 41% (Ekbom et al., 2005). The goal is to limit the increase to 20–25% (ETC, 2011).

An important objective for the development plant in Piteå has been to demonstrate stable and continuous operation. Development of the recovery boiler technology has led to increased availability. In the United States, Finland, Sweden, and Norway, the availability of recovery boilers is above 99.5%, and mill owners expect similar availability for gasification units. One of the most important design criteria for the recovery boiler is thus high availability. Since pulp production is a continuous process, it sets high standards for each process unit to work with almost no errors (Modig, 2005). The development plant in Piteå has been in operation for 12,000 h in total. The availability has gradually increased and during 2009 it reached approximately 70% on a monthly basis. However, the disruptions have been dominated by faults that would not have caused stops on a plant in full scale with better equipment (ETC, 2011). For example, Ekbom et al. (2005) assessed the potential of possible future full-scale "Nth" Black liquor gasification motor fuel (BLGMF) plants, and it is assumed that the mill invests in four gasifiers and that three gasifiers are in operation at any given time, with the fourth initially of interest on standby. As has been described, BLG was initially considered mainly as a technology for increasing power generation in the pulp and paper industry. In the beginning of the 2000s, however, Chemrec began to look at other possible usage of the syngas, including production of motor fuels, and during the last decade the focus has been shifted toward future implementation of BLGMF plants rather than BLGCC plants. However, the BLGCC concept could naturally still be interesting; future energy prices and policy instruments will determine the concept that is most profitable. The focus in the development of BLG has been, as described, on the gasification and gas cooling steps. The processes for cleaning and processing of the gas and synthesis of fuels such as methanol and DME are based on known, commercial technologies. However, for example, small differences in gas composition can be an important factor, and it is therefore important to demonstrate the whole process from black liquor to motor fuel. The two main goals of BLG II, which were achieved during the program period, were to be able to generate syngas that could be cleaned with known technologies and used for synthesis of methanol and DME, and to generate input data for upscaling to industrial scale (ETC, 2011).

In September 2010, a BLGMF demonstration plant was inaugurated in Piteå (Landälv et al., 2010). The plan was to start operation during Spring 2011. The plant converts the raw syngas from the gasifier, which was previously flared, into DME. The plant provides trucks that have been adjusted for operation on DME (designed by Volvo) with fuel for commercial test operation during approximately two years. The project has been partly financed by European Union and the Swedish Energy Agency (Landälv et al., 2010; ETC, 2011). Chemrec is also planning for a full-scale BLGMF plant at Domsjö Fabriker in Örnsköldsvik, Sweden. The goal is to start production of DME for use in heavy trucks and methanol for blending in gasoline in 2014. The final investment decision was planned (ETC, 2011). The project has received a grant from the Swedish Energy Agency that amounts to 500 MSEK, which has been approved by the European Commission. The total budget of the project is approximately 3000 MSEK. The Domsjö mill is a sulfite mill with two old boilers for energy and chemical recovery that are in need of replacement. The plant is to have $3 \times 50\%$ gasifier trains, each designed to gasify approximately 550 tDS per day (corresponding to approximately 100 MW). Thus, it is possible that the first full-scale plant for the Chemrec technology will be based on liquor from a sulfite mill and not black liquor from a kraft pulp mill, which has been the main fuel feedstock used in development of the technology. However, as mentioned, these liquors are similar. Based on the current situation for the development of BLG, it was assumed that large-scale implementation of full-scale BLG plants is unlikely to occur before around year 2020.

Naqvi et al. (2010) reviewed few interesting studies in Chemrec BLG system.

Marklund et al. (2007) identified most important parameters for a proposed computational fluid dynamics (CFD) modeling of Chemrec BLG as an initial step before a complete model validation against experimental data is done. The considered performance response parameters were

1. Fraction of volatile sulfur
2. Sulfide to sulfate ratio
3. Fraction of volatile carbon
4. CO to CO_2 ratio.

Results showed that the sensitivity to the amount of sulfur released to the gas phase as H_2S during devolatilization and concentration ratio of Na_2S and Na_2SO_4 in BLS have relatively large effects on performance response parameters as compared to carbon in volatile matter and CO/CO_2 concentration. The influential parameters appeared to be of great importance during model validation of CFD model against experimental data is considered.

Larson et al. (2006) and Larsson et al. (2006) investigated the inaccuracies in thermochemical data that influenced process variables resulting from equilibrium modeling of Chemrec oxygen-blown pressurized gasification process. The effect of the variation in pressure in the gasifier (25–32 bar) had a small effect on H_2S formation, but the data uncertainties became larger in temperature variations higher than

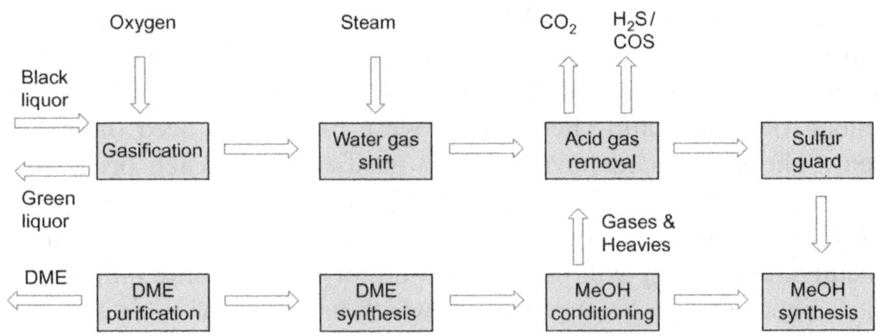

Figure 2.16 Schematic of the Piteå DME production plant. Reproduced with permission from Chemrec.

$1000°C$. A higher moisture content in both recovery system and gasifier favored $NaOH(g)$ and $KOH(g)$ formation. The impacts on sulfur equilibrium chemistry were found to be of utmost importance. The calculation results indicated significant uncertainties in $Na_2S_{(s,l)}$, $K_2S_{(s,l)}$, and $K_2CO_{3(s,l)}$ for $KOH_{(g)}$ formation.

The pressurized CHEMREC BLG systems offer the following benefits compared to recovery boilers.

- Effective pulping cooking chemicals recovery with simultaneous production of synthesis gas for high value-added green product streams
- Increased energy efficiency from utilization of syngas as feedstock for production of green electricity, automotive fuels, or hydrogen
- Dramatically improved green electricity yield through combined cycle power generation in the CHEMREC BLGCC system
- Addition of new green chemical products, methanol, or DME for automotive fuels from the CHEMREC BLGMF system
- Potential of novel production of green hydrogen for fuel cells and other uses from the CHEMREC BLGH2 system
- Option of utilizing new kraft cooking processes with higher yield
- New opportunity to manage the environmental impact of black liquor conversion
- No risk of smelt—water explosion.

Figures 2.15 and 2.16 show the new DME plant at Piteå and schematic of the Piteå DME production plant (Landälv, 2010) and Figure 2.17 shows BioDME trucks (Salomonsson, 2011). When used as a fuel in a diesel engine, BioDME gives an equally high efficiency

Figure 2.17 BioDME trucks are in operation today. Reproduced with permission from Volvo and Maria Faldt.

rating and a lower noise level compared with a conventional engine. Compared with diesel, BioDME generates an impressive 95% lower carbon dioxide emissions. What is more, its combustion produces very low emissions of particles and nitrogen oxides. This makes BioDME an ideal fuel for diesel engines. DME is a gas at ambient conditions but is liquefied at low pressure, just 5 bar and is handled as a liquid during distribution and use. It is simple to handle, similar to handling propane. DME can be produced from natural gas and also from various types of biomass. When it is produced from biomass, it is known as BioDME.

The BioDME project aims to demonstrate production of environmentally optimized synthetic biofuel from lignocellulosic biomass at an industrial scale (Landälv, 2010; Lindblom, 2012; Salomonsson, 2013). The project involves a consortium of Chemrec, Haldor Topsøe, Volvo, Preem, Total, Delphi, and ETC. The project is supported by the Swedish Energy Agency and the EU's Seventh Framework Programme. The final output of this demonstration is DME produced from black liquor through the production of clean synthesis gas and a final fuel synthesis step. In order to check technical standards, commercial possibilities, and engine compatibilities, BioDME will be tested in a fleet consisting of 14 Volvo trucks. In April 2012, Chemrec signed a cooperation agreement with China Tianchen Engineering Corporation, TCC, to provide an integrated offering of Chemrec plants on a global lump-sum turn-key basis and co-market the Chemrec BLG technology—a route to second generation biofuels or green power. The signing took place in Stockholm in the presence of the prime ministers of China, Wen Jiabao, and of Sweden, Fredrik Reinfeldt. Under the agreement, Chemrec and TCC will develop an offering to provide industry standard design, engineering, procurement, and construction (EPC) services as well as overall performance guarantees to support project financing for BLG plants. TCC will also assist in procuring plant financing.

The possible chemicals that can be produced from the syngas are hydrogen, methanol, DME, FT fuels, ethanol, and MTBE (Tampier et al., 2004).

The investment cost for a full-scaled pressurized BLG (PBLG) unit is estimated to be slightly higher than for a new conventional recovery boiler (Warnqvist et al., 2000). However, pressurized black liquor gasification with an integrated combined cycle (BLGCC) has the potential to double the amount of net electrical energy for a kraft pulp mill compared to a modern recovery boiler with a steam turbine (Axegård, 1999). For more closed systems with less need of steam, this increase in electrical energy will be even higher. Another advantage with the PBLG process is the increased control of the fate of sulfur and sodium in the process that can be used to improve the pulp yield and the quality for the mill. This control is very important for the green liquor quality and is quite limited with a conventional recovery boiler. A disadvantage with gasification is that it will increase the causticizing load. However, BLG has a lower requirement for make-up salt cake compared to the recovery boiler. Even though the PBLG process might have a lot of advantages compared to the recovery boiler there are still a number of uncertainties for this technology.

BLG is still a developing technology. It will take some time before reliable large units are available. BLG can produce more electricity (Vakkilainen et al., 2008). Current commercial atmospheric processes are not as energy efficient as the kraft recovery boiler process (Grace and Timmer, 1995; Mckeough, 2003). The black liquor gasifier needs to operate under pressure to have an electricity advantage.

Pressurized black liquor gasification can produce more electric power than recovery boiler processes (Table 2.5) (Vakkilainen et al. 2008; Warnqvist et al., 2000; Larson and Consonni, 2000; Larson et al., 2000). Pressurized black liquor gasification units cost more than utility-type recovery boilers. The revenue generated with extra electricity sales needs to be used to offset the additional investment. Producing extra electricity with BLG is only marginally competitive for integrated mills. But for large pulp mills, BLG approaches competitive figures (Paterson, 2003).

Even though there are significant gains to be made, there still remain many unresolved issues (Tucker, 2002; Katofsky et al., 2003): finding materials that survive in a gasifier, mitigating increased causticizing load, how to startup and shutdown, tar destruction, alkali removal, and achieving high reliability. The full impact of the BLG on recovery cycle chemistry needs to be carefully studied with commercial units. The first large demonstration units will cost two to three times more than a conventional recovery boiler. Although this will improve with time, price will hinder the progress of BLG. A small BLG with a commercial gas turbine size of 70 MWe requires a mill size of over 500,000 ADt/a. Commercial gasifiers probably need to be over 250 MWe in size. It is therefore expected that full-size black liquor gasifiers will be built in new greenfield mills, and not as replacement units of old recovery boilers.

Table 2.5 Power Balance with Gasification and Recovery Boiler Processes			
	Generation (kWh/ADt)	Generation (kWh/ADt)	Sales (kWh/ADt)
Pressurized gasification			
Chemrec type	2100	1050	1050
ABB type	2100	920	1180
New pulp mill with recovery boiler			
Conventional recovery boiler	1450	700	750
High power to heat recovery boiler	1700	730	970
Source: *Based on Vakkilainen et al. (2008), Warnqvist et al. (2000), Larson and Consonni (2000), Larson et al. (2000).*			

Clearly, BLG technology offers tremendous potential to make an impact on society. However, before it can totally replace the current recovery boiler technology, some work must be done to make it more economically attractive. One major area that requires attention is the causticization process. Gasification technology can cause significant increases in capacity for the lime cycle, requiring significant increases in fossil fuel consumption, and to improve economic viability, alternative causticization technologies must be considered.

Gasification is a well-established technique, but its application to black liquor is new and creates specific research needs. Perhaps the highest priority is to deal with the materials for constructing the gasifier. The process can operate at very high temperatures (up to 1000°C) and involves very aggressive molten salts (Na_2S, Na_2CO_3, $NaCl$) that tend to react strongly with ceramics and other materials. There is a very aggressive gas atmosphere (HCl, CO). This was an issue with the gasification system at Weyerhaeuser, New Bern. The problems were solved by using new materials and making some design changes (Brown et al., 2004). There are issues concerning the formation of tar and condensable organic matter. Approximately 1–5% of the carbon in black liquor is converted to methanol, ethanol, cresol, xylene, and a variety of other tar and condensable organic components. Several other questions need to be addressed. For example, can sodium and sulfur separation be controlled by process design or operation? How much H_2S is produced, rather than other sulfur-containing gases? And can H_2S be recovered efficiently from the product gases? Researchers around the world are trying to find answers.

2.2 CATALYTIC HYDROTHERMAL GASIFICATION OF BLACK LIQUOR

The process is also known as supercritical water oxidation. This is a novel technology to produce methane rich synthesis gas as an alternative to replace the conventional recovery system (Naqvi et al., 2010). Heat demand for bringing water to supercritical conditions is less than that for evaporating at subcritical pressure. This phenomenon leads to better energy savings as compared to conventional recovery system (Vogel et al., 2005; Sricharoenchaikul, 2009). High water content in the black liquor—nearly 80% under supercritical condition would increase gasification reactions that lead to high organics conversion to the synthesis gas (Calzavara et al., 2005; Williams and Onwudili, 2006). This phenomenon results in direct introduction of black liquor to hydrothermal gasifier removing energy demanding evaporation unit in conventional process, thus decreases the overall steam demand of the pulp mill. The multi-effect evaporation unit represents nearly 37% of total heat demand of the pulp mill and can be removed in this technique. Unlike other gasification processes, black liquor from the digestion unit is directly introduced to catalytic hydrothermal gasifier at supercritical water conditions (600°C, 300 bar). The catalytic hydrothermal gasification process involves three steps:

1. *Heat up phase* (decomposition unit): During this phase, larger molecules in black liquor hydrolyze to form coniferyl alcohols due to presence of lignin. Cellulose present in lignin decomposes rapidly in water at about 250°C.
2. *Salt separation*: In salt separator, inorganic salts present in black liquor precipitate and returned to the pulp mill.
3. *Catalytic reactor*: This is the actual methane synthesis unit where smaller organic molecules, such as carboxylic acid, alcohols, and aldehydes, are converted to CH_4, CO_2, H_2, and CO.

Carboxylic acid is converted to methane, carbon dioxide, and hydrogen. The tar formation is avoided due to supercritical conditions and presence of catalyst.

Savage et al. (1995) discussed the selection of suitable catalyst for methane production from organic waste streams at supercritical conditions such as ruthenium and activated carbon derived from coconut shells. Waldner and Vogel (2005) suggested stabilized Raney Ni catalyst for wet biomass hydrothermal gasification. Peterson et al. (2005) studied the design of salt separator used for biomass conversion to the synthesis gas but more research is required especially for black liquor. The application of catalysis in reactor is mainly used to lower the desired gasification temperature to get high conversion.

Sricharoenchaikul (2009) studied the feasibility of supercritical water oxidation technique to convert black liquor to value-added fuel products and recovery of pulping chemicals. The experiments were conducted in a quartz capillary heated in a fluidized bed reactor based on important operating parameters which were pressure, temperature, feed concentration, and reaction time. He found that pressure between

220 and 400 bar had insignificant influence on the synthesis gas and carbon conversion. But increasing temperature and residence time between 375–650°C and 5–120 s resulted in high carbon conversion, greater synthesis gas production, and energy efficiency.

REFERENCES

Ådahl, A., Harvey, S., Berntsson, T., 2004. Process industry energy retrofits: the importance of emission baselines for greenhouse gas reductions. Energy Policy 32, 1375–1388.

Andersson, E., Harvey, S., 2004. Pulp-mill integrated bio-refineries: a framework for assessing net CO_2 emission consequences. In: Proceedings, AIChE 2004 Fall Annual Meeting, Austin, TX, pp. 203–208.

Axegård, P., 1999. Kretsloppsanpassad massafabrik-Slutrapport, KAM 1 1996–1999, KAMrapport A31, Stiftelsen för Miljöstrategisk forskning.

Bajpai, P., 2008. Chemical Recovery in Pulp and Paper Making. PIRA International, UK, 166pp.

Bajpai, P., 2013. Biorefinery in the Pulp and Paper Industry. Elsevier Inc., UK, 114pp.

Bergek, A., 2002. Shaping and Exploiting Technological Opportunities: The Case of Renewable Energy Technology in Sweden (Ph.D. thesis). Department of Industrial Dynamics, Chalmers University of Technology, Göteborg, Sweden, 2002.

Berglin, N., Lindblom, M., Ekbom, T., 2002. Efficient production of methanol from biomass via black liquor gasification. In: Tappi Engineering Conference, San Diego, CA.

Björkman, A., "The Swedish new soda-recovery process project," Forum on Kraft Recovery Alternatives, 29–30 April 1976, Institute of Paper Chemistry, Appleton, Wisconsin, pp. 205–218 (1976).

Brown, C., Landälv, I., 2001. The Chemrec black liquor recovery technology—a status report. In: International Chemical Recovery Conference, Whistler, Canada, June 11–14, 2001.

Brown, C.A., Gorog, J.P., Leary, R., Abdullah, Z., 2004. The Chemrec black liquor gasifier at New Bern—a status report. In: International Chemical Recovery Conference, Charleston, SC, June 6–10, 2004.

Calzavara, Y., Joussot-Dubien, C., Boissonnet, G., Sarrade, S., 2005. Evaluation of biomass gasification in supercritical water process for hydrogen production. Energy Convers. Manage. 46, 615–631.

Chemrec, 2013. <http://www.Chemrec.se/>.

Connor, E., 2007.The integrated forest biorefinery: the pathway to our bio-future. In: International Chemical Recovery Conference: Efficiency and Energy Management, Quebec City, QC, Canada, May 29–June 1, 2007, pp. 323–327.

Copeland, G.G., 1969. The Copeland process fluid bed system and pollution control worldwide. In: Proceedings of the 1969 Purdue Industrial Waste Conference, pp. 1017–1023, 1969.

Copeland, G.G., Hanway, J.E., 1967. Fluidized bed oxidation of waste liquors resulting from the digestion of cellulosic materials for paper making. U.S. Patent No. 3,309,262.

Covey, G.H., Ostergren, M.E., 1985. DARS is the key to sulfur-free pulping. Pap. Trade J. 169 (5), 51–56.

Dahlquist, E., 2003. A combined physical and statistical simulation model for black liquor gasification. In: SIMS 2003 Conference in Vasteras.

Dahlquist, E., Jacobs, R., 1992. Development of a dry black liquor gasification process. In: Proceedings of the 1992 International Chemical Recovery Conference, Seattle, WA, pp. 457–471.

Dahlquist, E., Jacobs, R., 1994. Development of a dry black liquor gasification process. Pulp Paper Canada 95, 2.

Dahlquist, E., Jones, A., 2005. Presentation of a dry black liquor gasification process with direct caustization. TAPPI J. 15, 19.

Dahlquist, E., Avelin, A., Yan, J., 2009. Black liquor gasification in a CFB gasifier—system solutions. In: The First International Conference on Applied Energy (ICAE'09), Hong Kong, January 5–7, 2009.

Dance, M., 2005. Hydroxide Formation and Carbon Species Distributions During High-temperature Kraft Black Liquor Gasification (M.Sc. thesis). Georgia Institute of Technology, August 2005.

DeCarrera, R., 2002. Engineering study for a full scale demonstration of steam reforming black liquor gasification at Georgia-Pacific's mill in Big Island, Virginia. Final Report, Georgia-Pacific, Department of Energy, 2002, 24pp.

DeCarrera, R., 2006. Quarterly Technical Progress Report 20 Demonstration of Black Liquor Gasification at Big Island. Report 40850R20 <http://www.gp.com/containerboard/mills/big/pdf/rpt40850R20.pdf> (accessed 6.4.08).

Dehaas, G.G., 1979. Spent liquor treatment. U.S. Patent No. 4,135,968.

Dehaas, G.G., Hurley P.J., Root, D.F., 1976. Dry pyrolysis approach to chemical recovery. In: Forum on Kraft Recovery Alternatives, April 29–30, 1976. Institute of Paper Chemistry, Appleton, WI, pp. 103–123.

Dickinson, J.A., Verrill, C.L., Kitto, J.B., 1998. Development and evaluation of a low-temperature gasification process for chemical recovery from kraft black liquor. In: Proceedings of the 1998 International Chemical Recovery Conference, Tampa, FL, June 1–4, 1998.

Durai-Swami, K., Mansour, M., Warren, D., 1991. Pulsed combustion process for black liquor gasification. U.S. DOE Report DOE/CE/40893-T: (DE92003672).

ETC, 2011. Gasification of black liquor. Popular Science Report from the BLG II Program 2007–2010.

Ekbom, T., Lindblom, M., Berglin, N., Ahlvik, P., 2003. Technical and commercial feasibility study of black liquor gasification with methanol/DME production as motor fuels for automotive uses—BLGMF. Report for Contract No. 4.1030/Z/01-087/2001. European Commission, Altener Program, Stockholm, Sweden.

Ekbom, T. Berglin, N., Logdberg, S., 2005. BLG with motor fuel production – BLGMFII: a techno-economic feasibility study on catalytic Fischer–Tropsch synthesis for synthetic diesel production in comparison with methanol and DME as transport fuels, Nykomb Synergetics 2005.

Empie, H.J., 1991. Alternative kraft recovery processes. TAPPI J. 74 (5), 272–276.

Evans, J.C.W., 1975. Many process applications are appearing for fluid-bed reactors. Pulp Paper 49 (4), 82–85.

Grace, T.M., 1987. Chemical recovery technology—a review. In: Proceedings of China Paper'87 Conference, Shanghai, China, October 14–15, 1987.

Grace, T.M., Timmer, W.M., 1995. A comparison of alternative black liquor recovery technologies. In: Proceedings of the International Chemical Recovery Conference, Toronto, pp. B269–275.

Grigoray, O., 2009. Gasification of Black Liquor as a Way to Increase Power Production at Kraft Pulp Mills (Master thesis). Lappeenranta University of Technology Faculty of Technology Degree Program of Chemical Technology.

Harvey, S., Facchini, B., 2004. Predicting black liquor gasification combined cycle powerhouse performance accounting for off-design gas turbine operation. Appl. Thermal Eng. 24, 111–126.

Hess, H.V., Cole, E.L., 1971. Treatment of waste liquors from pulp and paper mills. U.S. Patent No. 3,558,426.

Hess, H.V., Cole, E.L., Franz, W.F., 1976. Coking of kraft pulping liquors at lowered pH. U.S. Patent No. 3,944,462.

Hess, H.V., Franz, W.F., Cole, E.L., 1978. Liquid phase coking of spent kraft pulping liquors, U.S. Patent No. 4,067,767.

Holme, G., 1976. "The NSP one stage process," New Recovery Processes for Black Liquor. Proceedings from the Gunnar Sundblad Seminar, 6–7 May 1976, Skövde, Sweden, pp. 139–145.

Holme, G.K., 1975. "Combustion of alkaline cooking liquor," U.S. Patent No. 3,867,251.

Horntvedt, E., 1968. The SCA-Billerud recovery process. In: Proceedings of the Symposium Recovery of Pulping Chemicals, Helsinki, Finland, May 13–17, 1968.

Horntvedt, E., 1976. SCA-Billerud pyrolysis process in kraft pulping, new recovery processes for black liquor. In: Proceedings from the Gunnar Sundblad Seminar, Skövde, Sweden, May 6–7, 1976, pp. 132–138.

Hurley, P.J., 1980. Energy balances for alternate kraft recovery systems. Chem. Eng. Progress 76 (2), 43–53.

Katofsky, R., Consonni, S., Larson, E.D., 2003. A cost–benefit analysis of black liquor gasification combined cycle systems. In: Proceedings of the TAPPI Fall Technical Conference: Engineering, Pulping & PCE&I, Chicago, 22pp.

Kenaf Black Liquor Gasification Study, 2006. EPRI, Palo Alto, CA and Tennessee Valley Authority, Muscle Shoals, AL, 1014515.

Kignell, J.E., 1989. Process for chemicals and energy recovery from waste liquors. U.S. Patent No. 4,808,264.

Kitto, J., 1997. Method for gasifying cellulosic waste liquor using an injector located within the bed of fluidized material. U.S. Patent No. 5,632,858.

Kulkarni, A.G., Mathur, R.M., Naithani, S., Pant, R., 1987. Present status of "DARS" technology and perspectives of its application to small pulp mills. IPPTA J. 24 (3), 71–79.

Landälv, I., 2010. Woods to wheels—Chemrec's BioDME demonstration plant at the Smurfit Kappa mill. Pulp Paper Int..

Landälv, I., Granberg, F., Nelving, H., 2010. Production of di methyl ether from kraft black liquor for use in heavy duty trucks. In: Proceedings of International Chemical Recovery Conference, Williamsburg, VA, March 29 – April 1, 2010, vol. 2, pp. 179–186.

Larson, E.D., Consonni, S., 2000. Preliminary economics of black liquor gasifier/gas turbine cogeneration at pulp and paper mills. J. Eng. Gas Turbines Power 122 (2), 255–261.

Larson, E., Consonni, S., Katofsky, R., 2006. A cost–benefit assessment of gasification-based biorefining in the kraft pulp and paper industry. Final Report, vol. 1., Princeton University and Politecnico di Milano.

Larson, E.D., McDonald, G.W., Yang, W., Frederick, W.J., Iisa, K., Kreutz, T.G., Malcolm, E. W., Brown, C.A., 2000. A cost–benefit assessment of BLGCC technology. TAPPI J. 83 (6), 1–15.

Larsson, A., Nordin, A., Backman, R., Warnqvist, B., Eriksson, G., 2006. Influence of black liquor variability, combustion, and gasification process variables and inaccuracies in thermochemical data on equilibrium modeling results. Energy Fuels 20, 359–363.

Lindblom, M., 2003. An overview of Chemrec process concepts. In: Sixth International Colloquium on Black Liquor Combustion and Gasification, Park City, UT, May 13–16, 2003.

Lindblom, M., 2012. Chemrec. History and Future Plants Chemrec Sales Presentation to ETF. Nenet <www.nenet.se/.../Chemrec_biodme_motor_fuels_from_the_forest_matsli...>.

MTCI, 1997. PulseEnhanced™ steam reforming and recovery of kraft spent liquor. Weyerhaeuser New Bern demonstration.

Magnusson, H., Warnqvist, B., 1980. The NSP project: an alternative to the conventional recovery furnace. Chem. Eng. Progress 76 (2), 54–56.

Maloney, J.D., 1976. An SCA-Billerud kraft chemical recovery process, Forum on Kraft Recovery Alternatives, 29–30 April 1976. Institute of Paper Chemistry, Appleton, WI, pp. 182–204.

Mansour, M.N., Steedman, W.G., Durai-Swamy, K., Kazares, R.E., Raman, T.V., 1992. Chemical and energy recovery from black liquor by steam reforming. In: International Chemical Recovery Conference, Seattle, WA, June 7–11, 1992.

Mansour, M.N., Durai-Swamy, K., Aghamohammadi, B., 1993. Pulsed combustion process for black liquor gasification. Second Annual Report U.S. DOE Report DOE/CE/40893-T2 (DE94002668).

Mansour, M.N., Durai-Swamy, K., Warren, D.W., 1997. Endothermic spent liquor recovery process. U.S. Patent No 5,637,192.

Marbe, Å., 2005. New Opportunities and System Consequences for Biomass Integrated Gasification Technology in CHP Applications (Ph.D. thesis). Department of Energy and Environment, Division of Heat and Power Technology, Chalmers University of Technology, Göteborg, Sweden, 2005.

Marklund, M., 2006. Pressurized Entrained-flow High Temperature Black Liquor Gasification CFD based Reactor Scale-up Method and Spray Burner Characterization (Ph.D. thesis). Luleå University of Technology.

Marklund, M., Tegman, R., Gebart, R., 2007. CFD modelling of BLG: identification of important model parameters. Fuel 86, 1918–1926 (ISSN 0016-2361).

Maunsbach, K., Isaksson, A., Yan, J., Svedberg, G., Eidensten, L., 2001. Integration of advanced gas turbines in pulp and paper mills for increased power generation. J. Eng. Gas Turbines Power Trans. ASME 123 (4), 734–741.

Mcilroy, R.A., Kuchner, R.A., Monacelli, J.E., Johnson, D.W., 1997. Black liquor gasifier. U.S. Patent No. 5,645,616.

Mckeough, P., 1993. Research on black liquor conversion at the technical research center of Finland. Bioresour. Technol. 46 (1–2), 135–143.

Mckeough, P., 2003. Evaluation of potential improvements to BLG technology. In: Colloquium of Black Liquor Combustion and Gasification, Park City, UT, 12pp.

Middleton, T., 2006. Steam reforming technology at the Norampac Trenton mill. In: Presentation at IEA Meeting, Annex XV Black Liquor Gasification, Washington, NC, February 20–22, 2006.

Modig, G., 2005. Black Liquor Gasification: An Assessment from the Perspective of the Pulp and Paper Industry (Licentiate thesis). Environmental and Energy Systems Studies, Lund University, Lund, Sweden, 2005.

Möllersten, K., Yan, J., Westermark, M., 2003a. Potential and cost–effectiveness of CO_2-reduction in the Swedish pulp and paper sector. Energy 28, 691–710.

Möllersten, K., Yan, J., Moreira, J., 2003b. Potential market niches for biomass energy with CO_2 capture and storage. Biomass Bioenergy 25 (3), 273–285.

Möllersten, K., Lin, G., Yan, J., Obersteiner, M., 2004. Efficient energy systems with CO_2 capture and storage from renewable biomass in pulp and paper mills. Renewable Energy 29, 1583–1598.

Myers, R.L., Miller, R.L., 1976. St. Regis hydropyrolysis process, Forum on Kraft Recovery Alternatives, 29–30 April 1976. Institute of Paper Chemistry, Appleton, WI, pp. 74–102.

Naqvi, M., Yan, J., Dahlquist, E., 2010. Black liquor gasification integrated in pulp and paper mills: a critical review. Bioresour. Technol. 101, 8001–8015.

Newport, D.G., Rockvam, L., Rowbotton, R., 2004. Black liquor steam reformer start-up at Norainpac. In: Proceedings of TAPPI International Chemical Recovery Conference, South Carolina.

Nohlgren, I., 2004. Non-conventional caustization technologies: a review. Nordic Pulp Pap. Res. J. 19 (4), 467–477.

Nohlgren, I., Sinquefield, S., 2004. Black liquor gasification with direct causticization using titanates: equilibrium calculations. Ind. Eng. Chem. Res. 43 (19), 5996–6000.

Paterson, M. 2003. Evaluation of potential improvements to BLG technology. In: Colloquium of Black Liquor Combustion and Gasification, Park City, UT, May 13–16, 2003, 12pp.

Patrick, K., Siedel, B., 2003. Gasification edges closer to commercial reality with three new N.A. mills Startups. PaperAge, October 2003.

Peterson, A., Maurice, H., Waldner, M.H., Vogel, F., Jefferson, W., 2005. Fuels from biomass: use of neutron radiography to improve the design of a salt separator in supercritical-water biomass gasification. Massachusetts Institute of Technology, Cambridge.

Pettersson, K., 2011. Black Liquor Gasification-based Biorefineries—Determining Factors for Economic Performance and CO_2 Emission Balances Heat and Power Technology (Ph.D. thesis). Department of Energy and Environment, Chalmers University of Technology, Göteborg, Sweden, 2011.

Rao, N.J., Kumar, R., 1987. Ferrite recovery process—a promising alternate for small paper mills. IPPTA J. 24 (3), 30–37.

Richards, T., Nohlgren, I., Warnqvist, B., Theliander, H., 2002. Mass and energy balances for a conventional recovery cycle and for a cycle using borate or titanates. Nordic Pulp Pap. Res. J. 17 (3), 213–222.

Rockvam, L., 2001. Black liquor steam reforming and recovery commercialization. In: International Chemical Recovery Conference, Whistler, Canada.

Salomonsson, P., 2011. BIODME project. In: Seventh Asian DME Conference, Nigata, Japan.

Salomonsson, P., 2013. Final report of the European BioDME project. In: Fifth International DME Conference, Ann Arbor, April 18, 2013.

Savage, P., Martino, C., Brock, E., 1995. Reactions at supercritical conditions, application and fundamentals. AIChE J. 41 (7), 1723–1778.

Saviharju, K., 1993. Black liquor gasification: results from laboratory research and test rigs. Bioresour. Technol. 46 (1–2), 145–151.

Sinquefield, S., 2005. In situ causticizing for black liquor gasification. Phase 2 Tropical Report, February 1, 2004–October 31, 2005.

Sricharoenchaikul, V., 2001. Fate of carbon containing compounds from gasification of kraft black liquor with subsequent catalytic conditioning of condensable organics, PhD dissertation, GA Institute of Technology, Atlanta 2001.

Sricharoenchaikul, V., 2009. Assessment of black liquor gasification in supercritical water. Bioresour. Technol. 100, 638–643.

Stevens, D.J., 2001. Hot gas conditioning: recent progress with larger-scale biomass gasification systems. Update and summary of recent progress. NREL/SR-510-29952.

Stigsson, L., 1998. Chemrec black liquor gasification. In: International Chemical Recovery Conference, Tampa, FL, June 1–4, 1998.

Suresh, P., 2002. Biomass Gasification for Hydrogen Production—Process Description and Research Needs. IEA Thermal Gasification Task Leader Gas Technology Institute, IL.

Swedish Energy Agency, 2008. System Aspects of Black Liquor Gasification—A Review of Existing Reports. ISSN 1403-1892.

Tadashi, N., Saisei, M., Noriyoshi, N., 1976. Method of recovering sodium hydroxide from sulfur free pulping or bleaching waste liquor by mixing ferric oxide with condensed waste liquor prior to burning. US Patent No. 4,000,264.

Tampier, M., Smith, D., Bibeau, E., Beauchemin, P.A., 2004. Identifying Environmentally Preferable Uses for Biomass Resources—Stage 1 Report: Identification of Feedstock-to-Product Threads. Report. Envirochem Services Inc., North Vancouver, BC.

Timpe, W.G., 1973.Pyrolysis of spent pulping liquors. U.S. Patent No. 3,762,989, 1973.

Timpe, W.G., Evers, W.J., 1973. The hydropyrolysis recovery process. TAPPI J. 56 (8), 100–103.

TRI Biomass Gasification Technology—How It Works, 2013. <http://www.tri-inc.net/pdfs/TRI% 20How%20It%20Works%20Overview.pdf>.

Tucker, P., 2002. Changing the balance of power. Solutions 85 (2), 34–38.

US DOE, 1997. High-solids black liquor firing in pulp and paper industry kraft recovery boilers. Phase 1a: Final report. Low temperature black liquor gasifier evaluation. U.S. DOE Report DOE/ER/10002-T1.

Vakkilainen, E.K., Kankkonen, S., Suutela, J., 2008. Advanced efficiency options: increasing electricity generating potential from pulp mills. Pulp Paper Canada 109 (4), 14–18.

Venkoba Rao, G., 1987. Direct alkali recovery system (DARS)—the new state of art system—a review. IPPTA J. 24 (3), 20–28.

Vogel, F., DiNaro, J., Marrone, P., Rice, S., Peters, W., Smith, K., Tester, J., 2005. Critical review of kinetic data for the oxidation of methanol in supercritical water. J. Supercrit. Fluids 34, 249–286.

Waldner, M., Vogel, F., 2005. Renewable production of methane from woody biomass by catalytic hydrothermal gasification. Ind. Eng. Chem. Res. 44 (13), 4543–4551.

Warnqvist, B., Delin, L., Theliander, H., Nohlgren, I., 2000. Teknisk ekonomisk utvärdering avsvartlutförgasnings processer. Värmeforsk Service AB, Stockholm.

Warnqvist, B., Delin, L., Theliander, H., Nohlgren, I., 2001. Techno-economical evaluation of black liquor gasification processes. In: International Chemical Recovery Conference, Whistler, BC, Canada.

Watkins, J.J., Timpe, W.G., 1980. Pyrolysis of spent pulping liquors. U.S. Patent No. 4,208,245.

Whitty, K., 2005. Black liquor gasification: development and commercialization update. In: ACERC Annual Conference, Utah.

Whitty, K., 2009. The changing scope of black liquor gasification. University of Utah, Salt Lake City.

Whitty, K., Baxter, L., 2001. State of the art in black liquor gasification technology. In: Joint International Combustion Symposium, Kauai, Hawaii, September 9–12, 2001.

Whitty, K., Nilsson, A., 2001. Experience from a high temperature, pressurized black liquor gasification pilot plant. In: Proceedings of International Chemical Recovery Conference, Whistler, BC, Canada, 2001, pp. 281–287.

Whitty, K., Verrill, C.L., 2004. A historical look at the development of alternative black liquor recovery technologies and the evolution of black liquor gasifier designs. In: International Chemical Recovery Conference, Charleston, SC, June 6–10, 2004.

Williams, P., Onwudili, J., 2006. Subcritical and supercritical water gasification of cellulose, starch, glucose, and biomass waste. Energy Fuels 20, 1259–1265.

Zeng, L., van Heiningen, A.. 1996. In: Proceedings of 82nd Annual Conference, Technology Section, CPPA, Montreal, 1996, A259.

Zou, X., Avedisian, M., van Heiningen, A., 1992. Kinetics of the direct causticization reaction between sodium carbonate and titanium dioxide. In: AIChE Forest Products Symposium Series. AIChE, New York.

Market Opportunities

The simplest application of BLG is to replace the boiler with a gasifier. One alternative is to build a greenfield mill, and the other is to replace a worn-out boiler with a gasifier. New mills may be built in Europe, but consensus holds that few new mills will be built in North America also. So it is expected that the paths to commercialization will differ between Europe and North America. In Europe, BLG is likely to converge to one or two well-defined biorefinery models. The United States is likely to have more BLG paths than in Europe, but they are not necessarily more profitable (Farmer, 2005; Farmer and Sinquefield, 2007, 2009). To understand how commercialization is likely to appear in different areas, the basic economic advantage of BLG should be remembered. The first issue is biomass transport which is an economic issue for all bioenergy. If pulp making can justify the management costs of silviculture plus the costs to collect and ship a bulky biomass to a pulp mill, it will remove the primary economic obstacle to bioenergy production—the prohibitive expense of biomass transport—and this is no mean accomplishment. Most biofuel efforts involve moving huge amounts of biomass to a processing center for the sole purpose of energy production, such as switch grass to diesel and corn stalks to ethanol. All move biomass only to acquire energy. So even if the economic returns to pulp making are quite low relative to the energy products or biorefinery products created, joint production will typically require some pulp making to make BLG economically attractive, if only to deliver the biomass.

Another concern is diversification of product lines. A kraft mill that adopts BLG exclusively to displace the boiler and to produce more electricity can be profitable (Larson and Raymond, 1997; Larson et al., 2003, 2006). But at present, a much sharper biorefinery with a diverse product range adds many new options that may become economically attractive if BLG technology can solve a few technical problems (Bajpai, 2008). Since the investigation of Grace and Timmer (1995) and Whitty and Baxter (2001), what has excited new interest in

BLG is the ability to leverage BLG into a wider portfolio of products, retrofitted to different types of mill, including medium-sized mills facing shutdown. The Chemrec development plant in Piteå, Sweden, is the most developed BLG model with strong industry momentum. Volvo Technology Transfer AB and affiliates now hold a majority share. And perhaps the best BLG path in Europe is the Chemrec R&D plan (Landälv, 2006). The biorefinery aims to produce pulp and to use the BLG syngas to produce DME fuel. DME is clean-burning and has advantages over diesel. Chemrec has implemented a DME demonstration in Piteå.

The Chemrec 2020 plan anticipates five large and three small BLGMF plants. Economic projections are attractive. This is largely due to returns from DME (Ekbom et al., 2003, 2005). Official commercialization of a Chemrec gasifier is virtually the model approach to BLG commercialization in Europe. Given the investment by Volvo Technology Transfer AB, there is relatively strong support to make DME fuels for the heavy vehicle industry and there is a move by Volvo to adapt its heavy vehicles so they can work better with DME fuels if the technology takes off. Europe's mills are more homogeneous than North America and its newer mills are built on an economical scale, so if corrosion and causticizing can be resolved or mitigated, this biorefinery model of BLG could be successfully commercialized within 10 years and reach full stride in 15 years.

Compared to Europe, North America has a different market with different mill infrastructure. Its mills are smaller and have difficulty achieving proper economies of scale for electricity production and pulp production. The US energy economy is unlikely to absorb a geographically diffuse DME market. Shifting the truck and car fleet at US motor companies, similar to the shift by Volvo, is likely to be the easier part of adjusting the infrastructure. DME distribution for long-haul trucking could be more formidable. Moreover, the mill infrastructure in North America is significantly older. This allows competition from Brazil, China, and Indonesia to remain profitable with a kraft process that uses a conventional recovery boiler, because their mills are newer, operate on a larger scale, have lower labor costs, and enjoy rapid tree rotation. These countries are good candidates for BLG adoption, but they are unlikely to be among the first adopters. This leaves the strategic adjustment to BLG in North America in a difficult position. Mills with a limited capital base will have to innovate first and invest faster in a new, capital-heavy technology that may not have been fully vetted before adoption. Moreover, diversity among the mills may mean there are fewer off-the-shelf models that can be replicated as mills adopt several niche applications of BLG. Mills can either generate one very high-value energy product in a simple boiler replacement model and hope for the best, or consider a much more technically complex, integrated biorefinery with a much more extensive product line of new specialty products that have high value or high growth potential.

According to IEA Bioenergy Report (2009), there are 236 recovery boilers in the world that have not been rebuilt during the last 20 years and thus can be suitable for replacement with gasification technology. However, the majority of these boilers have quite low capacities, $<500-600$ tDS per day. A BLGMF system would not be a realistic replacement alternative for these small boilers. One can assume that a mill which is replacing an outdated recovery boiler would desire somewhat more capacity (perhaps 25%) than the old boiler provided. A BLGMF system is a competitive alternative for capacities of roughly 1000 tDS per day and higher. Hence, the actual market is for replacement of boilers with a capacity of 800 tDS per day or more, and which have not been built or extensively renovated in the last 20 years. There are 57 such boilers in the world today, about half of which are in the United States. The majority of the remainder are located either in Canada or Japan. The market for the BLGMF system will expand in the future due to the obsolescence of more and larger recovery boilers. In short, each of the world's 327 recovery boilers with a capacity of more than 800 tDS per day can be considered a candidate for eventual replacement by a BLGMF system. It is becoming common for mills with multiple recovery boilers to replace several or all with one unit which has a capacity of 2000 tDS per day or more. A BLGMF system is clearly an alternative for these mills, so the market is actually larger than earlier suggested. This is shown in Figure 3.1 which presents the year of start-up and rebuild of North American recovery boilers.

Chemrec has sold its BLG plant in Piteå in northern Sweden and DME synthesis plant to Luleå University of Technology (LTU) to be used in continued R&D work. The pressurized, oxygen-blown gasifier of this plant has till date been operated more than 18,000 h, consistently producing syngas of very good quality. Since 2011, the syngas produced has been used for synthesis of biomethanol and bioDME.

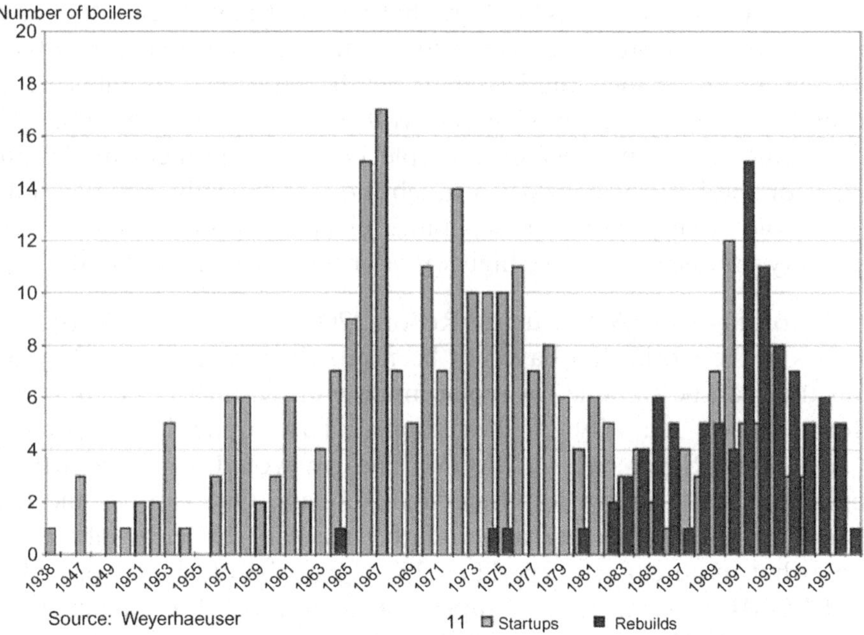

Figure 3.1 Year of start-up and rebuild of North American recovery boilers. Reproduced with permission from IEA Bioenergy Report (2009).

The bioDME produced has been used in very successful heavy truck fleet trials conducted by Volvo trucks within the pan-European BioDME project. Overall, this extended period of operation has validated the Chemrec gasification concept and provided all information required for commercial-scale implementation. Effective from December 31, 2012, these plants have been transferred to LTU. Also the operating and development personnel at the plant are now employed by LTU. In December 2012, the Swedish Energy Agency approved funding supporting the continued operation of the plants to make it available for new R&D programs. Chemrec actively participates in these programs and has retained the right of access to the plant for future trials, e.g., with feedstock of specific customers. Chemrec in its new form will thus for further development work rely largely on this new network structure and the core organization will be heavily focused on the commercialization of the technology (Chemrec, 2013).

REFERENCES

Bajpai, P., 2008. Chemical Recovery in Pulp and Paper Making. PIRA International, UK, 166pp.

Chemrec, 2013. <http://www.Chemrec.se/>.

Ekbom, T., Lindblom, M., Berglin, N., Ahlkiv, P., 2003. Technical and commercial feasibility study of BLG with methanol/DME production as motor fuels for automotive uses—BLGMF. Altener.

Ekbom, T., Berglin, N., Logdberg, S., 2005. BLG with motor fuel production—BLGMFII: a techno-economic feasibility study on catalytic Fischer—Tropsch synthesis for synthetic diesel production in comparison with methanol and DME as transport fuels. Nykomb Synergetics.

Farmer, M.C., 2005. Adaptable biorefinery: some basic economic concepts to guide research selection. In: Tappi Engineering, Pulping and Environmental Conference, Philadelphia, PA, 2005.

Farmer, M.C., Sinquefield, S., 2007. A biomass inventory of supplemental agricultural residuals within ten miles of pulp mills in the southeastern US. Texas Technical University.

Farmer, M.C., Sinquefield, S., 2009. Black Liquor Gasification: Paths and Obstacles to Commercialisation. Pira Industry Insight, UK.

Grace, T.M., Timmer, W.M., 1995. A comparison of alternative black liquor technologies. In: International Chemical Recovery Conference, Toronto, 1995.

IEA Bioenergy Report, 2009. Black liquor gasification summary and conclusions from the IEA Bioenergy ExCo54 Workshop.

Landälv, I., 2006. BLG: experiences at Piteå and New Bern and the pulp mill biorefinery. In: Chemical Pulping and Biorefinery Conference, Stockholm.

Larson, E.D., Raymond, D.R., 1997. Commercializing black liquor and biomass gasifier/gas turbine technology. TAPPI J. 12, 50.

Larson, E.D., Consonni, S., Katofsky, R.E., 2003. A cost—benefit assessment of biomass gasification power generation in the pulp and paper industry. Princeton Environmental Institute. Available at: <www.princeton.edu/~energy>.

Larson, E.D., Consonni, S., Katofsky, R.E., Iisa, K., Frederick, W.J., 2006. A cost benefit assessment of gasification-based biorefining in the kraft pulp and paper industry. Final Report 2006. Available at: <www.princeton.edu/~energy/publications/index.html>.

Whitty, K., Baxter, L., 2001. State of the art in BLG technology. In: Joint International Combustion Symposium, Kauai, 2001.

CHAPTER 4

Obstacles to Implementation of Black Liquor Gasification

BLG is a promising technology for the pulp and paper industry but there are obstacles hindering its commercialization (Farmer and Sinquefield, 2009; IEA Bioenergy report, 2009; Pettersson, 2011) (Table 4.1).

The most important of them are described below (Grigoray, 2009; Farmer and Sinquefield, 2009).

4.1 FINANCIAL RISKS

The pulp and paper industry is a highly capital intensive industry. The recovery system is the biggest investment at a pulp mill. Literature shows that the installed cost of a BLGCC will go beyond the cost of an existing recovery system by two times approximately. Kreutz et al. (1998) have reported that the capital costs of a Tomlinson-based system calculated for pulp mills having the productivity of 1580 dry short tons of unbleached pulp per day is \$136.2 million and about \$218 million for HTBLGCC (high temperature black liquor gasification combined cycle). Larson (2006) has reported that the operating and maintenance costs of gasification technology will be higher than for traditional recovery. So, it is clear that the gasification system demands significant capital investment. Mills need guarantees that the process will be successful on an industrial scale before making investment. The trials of the BLG given in the public is not enough, because

Table 4.1 Technical Obstacles to BLG

Obstacles	Seriousness	Likely Solutions	Possibility of Success	Number of Years Before a Solution
Material corrosion and stress	Critical	New gasifier metals, refractories, nonsulfur pulping	Moderate, progress will come soon	7 then ongoing
Tar condensation	Critical	Tar cracker catalyst	Moderate to good	3–5
Causticizing	Serious	*In situ* causticizing; 100% for titanates	Moderate, enough to prevent lime kiln redesign	5
Hot gas cleanup	Serious	New cleanup technologies	Low	7–10
Steam deficit	Mild	Supplemental biomass shipped to mill	Moderate, new biomass field process technology	5

the behaviors of this technology in the combined cycle and problems which may arise remain unclear. The countries where renewably energy sources play a great role in their energy sector are ready to make investment in the gasification of black liquor as a way to increase power capacity. In Finland, the interest shown to this process and associated risks are justified because this country is the leader of bioenergy production having extensive knowledge in this field and where more than 10% of energy is based on biomass (Finnish Forest Industries, 2009). In countries like Russia, where the share of renewable energy sources used, excluding hydro power, is approximately 0.5% and the amount of nonrenewable sources is significant, the country will not accept active participation in the development of BLG technology without more serious reasons than increasing production of electric energy (Vetrov, 2009).

4.2 PROTECTION OF THE GASIFIER

The high temperature and pressure and the alkaline conditions create aggressive environment. Protection from an aggressive environment is very important in the operation of a gasifier because it determines the lifetime and hence the feasibility of gasification as a technology as a whole. There are two types of gasifier using different types of shell protection (Finnish Forest Industries, 2009):

1. Refractory brick lining
2. Refractory-coated coiled lining.

The gasifier installed in Stoghall has the refractory brick lining, or the fusion-cast alumina refractory to be precisely, which is identical to one installed at New Bern's booster after its reconstruction. After 1000 h of work, the gasifier was shutdown and the structure of refractory bricks was studied. The results showed that at the top of the bricks, the same products of reactions between the bricks and smelt as in case of the booster's refractory were found, characterized by the partial destruction of the lining. The service life of a gasifier with such a refractory lining is approximately one year. Further studies were carried out on increasing the operation period of the gasifier. The studies showed that it is possible to extend this period to more than two if fusion-cast magnesia–alumina spinel refractory is used. In contrast, two essential shortcomings of fusion-cast based refractory are high costs and thermal shock sensitivity (Keiser, 2003, 2009). The second type of high temperature high pressure gasification (HTHP) gasifier has cooling coil screens representing refractory-coated helically coiled metal tubes, in which the cold water, under pressure, is intended to cool refractory surface below the melting point of smelt and to reduce the thermal shock to the gasifier's shell (Hemrick, 2004). The material used as a protective layer of the tubes should have a coefficient of thermal expansion which allows it to stay in good contact with the tubes besides good thermal conductivity and corrosion resistance. Several tests were conducted to select coatings with high adhesion, surface quality, and resistance to smelt. Specifically, alumina and magnesia cement, alumina ram mix, magnesia castable were put through trials. Among them, magnesia castable showed better results (Hemrick, 2004). A material which provides high quality and economic protection of the gasifier has not been found yet for both designs of the gasifier. Research in this area is needed.

4.3 INCREASE IN THE CAUSTICIZING DEMAND

Increase in the causticizing demand is another obstacle of high- and low temperature gasification (Farmer and Sinquefield, 2009). In a traditional type of recovery boiler, the reducing conditions in the char bed region produce some causticizing of the smelt before it leaves the boiler. Furthermore, a portion of the sodium in the smelt is bound up as sodium sulfide, thus reducing the amount of sodium available to form sodium carbonate. Causticizing of the remaining sodium carbonate is subsequently completed by the lime cycle. In LTBLG, most of the sulfur ends up in the gas phase as hydrogen sulfide, and the solid product is mainly creating an even greater causticizing load than for conventional recovery. If scrubbing is used to remove hydrogen sulfide from the gas, then some carbon dioxide will be coabsorbed which must also be causticized. The inherent separation of sulfur and sodium creates opportunities for high-yield pulping but the additional causticizing load must be tackled. For HTBLG, the situation is to some extent better. The sulfur is split between the gas and solid phases so a smaller portion of the sulfur is available to bind up sodium as sodium sulfide in comparison to conventional recovery. In the Chemrec HTBLG system, the water quenching of the smelt results in coabsorption of carbon dioxide, which must then be causticized. While the reasons are different for each case, increased causticizing load will be a problem for HTBLG and LTBLG. The increase can range from 15% for HTBLG to 50% or more for LTBLG. If an existing mill has excess lime cycle capacity, then the extra caustic load from BLG may not be much of a problem beyond the extra energy used. If not, then the extra expense of increasing caustic plant capacity could well deter the mill from implementing BLG.

Direct causticizing, also called autocausticizing, is a likely solution that has received significant attention (Bajpai, 2008). It involves adding a chemical agent to the liquor that preferentially binds with sodium to displace carbonate and force it into the gas phase as carbon dioxide. The resulting complex then releases sodium during the hydrolysis step to produce sodium hydroxide and return the agent to its original form. There are two situations:

1. To mitigate the increase in caustic load (partial causticizing) so as to allow an existing lime cycle to continue service
2. To accomplish 100% causticizing during gasification and thereby eliminating the need for the costly lime cycle.

Several chemical agents have been identified (Nohlgren, 2004). Borates, titanates, and manganates have been found to be very effective. Borate autocausticizing has been used in recovery boilers to increase capacity. The borate addition generates caustic in green liquor at the recovery boiler smelt dissolving tank. The caustic present in green liquor then reduces the amount of lime required for causticizing. Borates remain soluble and circulate through the pulping cycle and the lime cycle. The reaction occurs in the gasifier while the organics are being gasified:

$$NaBO_2 + Na_2CO_3 \leftrightarrow Na_3BO_3 + CO_2$$

The caustic is then recovered after hydrolysis:

$$Na_3BO_3 + H_2O \leftrightarrow 2NaOH_{(aq)} + NaBO_{2(aq)}$$

The research with borates for BLG has yielded varied results. Sinquefield et al. (2007) reported that borates were effective for 20% causticizing during HTBLG but the pressure was limited to 5 bar (0.5 MPa), not high enough for the pressurized HTBLG process, and even then the lime cycle is still required. Nohlgren and Sinquefield (2007) found borates ineffective for 25% causticizing of a kraft liquor, but found satisfactory results with a soda liquor. Sodium trititanate is added to the liquor to bind up the sodium as pentatitanate and allow the carbon to be released as carbon monoxide or carbon dioxide. The insoluble sodium trititanate is recycled back to the black liquor entering the gasifier. The basic chemistry is given below:

$$7Na_2CO_3 + 5(Na_2O \cdot 3TiO_2)_{(s)} \leftrightarrow 3(4Na_2O \cdot 5TiO_2)_{(s)} + 7CO_2$$

The sodium pentatitanate formed in the gasifier is leached with water forming solid sodium trititanate and sodium hydroxide (white liquor):

$$3(4Na_2O \cdot 5TiO_2)_{(s)} + 7H_2O \leftrightarrow 14NaOH_{(aq)} + 5(Na_2O \cdot 3TiO_2)_{(s)}$$

Sinquefield et al. (2007) reported that titanates were effective for complete causticizing, but only for HTBLG the absolute pressure was limited to 5 bar (0.5 MPa). Higher pressures yielded high CO_2 partial pressures, which inhibited the conversion. Although titanates did not work for the specified LTBLG conditions, i.e., 600°C, thermodynamic modeling indicated that causticizing should take place above 650°C. Manganese(II,III) oxide is added to bind up the sodium and allow carbon to be released as CO or CO_2:

$$Na_2CO_3 + Mn_3O_4 \leftrightarrow 2NaMnO_2 + MnO + CO_2$$

Manganese(II,III) oxide is recovered and hydroxide is formed upon hydrolysis:

$$2NaMnO_2 + MnO + H_2O \rightarrow 2NaOH_{(aq)} + Mn_3O_{4(s)}$$

Initial tests with BLG found manganates to be effective for 100% causticizing in LTBLG.

However, potassium in the liquor bound with the manganate and was not released on hydrolysis. Thus the agent would not return to its original form for cyclic chemistry. No causticizing agent was found to work with LTBLG.

4.4 TAR CONDENSATION

Tars are heavy organic compounds. Many tar species are known carcinogens. These result from incomplete gasification of the lignin and also due to other high molecular weight compounds in the black liquor. When gasified, all fuels are found to produce some tar. Tars account for 0.1% of the organic carbon in the exiting syngas in a well-running gasifier and in the worst cases, 20% of the carbon can form tars. Tars fare also found to foul processing equipment if the gases are cooled below about 200–250°C. The best way to get rid of tars is not to form them, which means having good gas/solid contact and sufficient residence time for gasification. If the syngas is to be used for high value processes, such as firing a turbine or formation of alcohols or FT diesel, then tar compounds must be removed from the syngas stream. Best results have been obtained with catalytic oxidation. Many commercial catalysts will increase the oxidation of tar species, but most are poisoned by sulfur gases—H_2S, COS. Sinquefield et al. (2007) found a catalyst that is stable in the presence of sulfur and was tested with a real BLG syngas. The catalyst has yet to be pilot tested. Sricharoenchaikul (2001) studied several commercial catalysts and found only one that was not quickly poisoned by H_2S; it was slowly deactivated and a reactivation process was developed.

4.5 HOT GAS CLEANUP

Syngas must be cleaned of impurities if it is to be used for high value products. These impurities are particulates, sulfur gases, chlorine gases, and tars. It is a costly affair to cool down the gas for classic scrubbing and filtering and then reheat it. Sometimes it even has to be repressurized. This problem is not exclusive to BLG. Syngas from any fuel source needs to be cleaned before high value use, such as sulfur from coal gasification. Large-scale operations typically have feasible economics where it is needed. The ability to clean impurities from hot pressurized gas will save substantial costs for all forms of gasification. With sorbents, sulfur and acid gases can be removed without cooling. The sulfur gases can then be recovered when the sorbant is reactivated. Tars can be catalytically destroyed. Membrane technology is improving. State-of-the-art membranes can bear a temperature of up to 250°C and higher, not genuine hot cleanup but better than wet scrubbing. Improvements are required in high temperature pressurized removal of fine particulates, ammonia, chlorine, and light hydrocarbons. The requirements depend on the purpose of the syngas. The syngas is cleaner if the application is more valuable.

4.6 STEAM DEFICIT

The combustion energy from black liquor is used to make steam. This is used in the papermaking process. Gasification usually offers higher thermal efficiency but if the syngas is to be used for motor fuels, then extra fuel must be brought in to produce the lost steam. It is presumed this will be in the form of bark or agricultural biomass. The size of the steam deficit depends on the purpose of the syngas, but the matter needs to be taken into consideration.

REFERENCES

Bajpai, P., 2008. Chemical Recovery in Pulp and Paper Making. PIRA International, UK, 166pp.

Farmer, M.C., Sinquefield, S., 2009. Black Liquor Gasification: Paths and Obstacles to Commercialisation. Pira Industry Insight, UK.

Finnish Forest Industries, 2009. <http://www.corporateregister.com/a10723/ffif06-env-fin.pdf>.

Grigoray, O., 2009. Gasification of Black Liquor as a Way to Increase Power Production at Kraft Pulp Mills (Master thesis). Lappeenranta University of Technology Faculty of Technology Degree Program of Chemical Technology.

Hemrick, J.G., 2004. Refractory testing and evaluation at Oak Ridge National Laboratory for black liquor gasifier application. Refract. Appl. News 9 (6), 20 (Oak Ridge National Laboratory, Oak Ridge, TN).

IEA Bioenergy Report, 2009. Black liquor gasification summary and conclusions from the IEA Bioenergy ExCo54 Workshop.

Keiser, J.R., 2003. Corrosion issues in black liquor gasifier. In: Colloquium on Black liquor Combustion and Gasification, Salt Lake City, May 13–16, 2003. Oak Ridge National Laboratory, Oak Ridge, TN.

Keiser, J.R., 2009. Improved materials for high-temperature black liquor gasification. Final Technical Report, June 2006, 39pp.

Kreutz, T., Consonni, S., Larson, E., 1998. Performance and preliminary economics of black liquor gasification combined cycles for a range of kraft pulp mill sizes. In: Tappi International Chemical Recovery Conference, Tampa, FL, June 1–4, 1998.

Larson, E., 2006. A cost–benefit assessment of gasification-based biorefining in the kraft pulp and paper industry. Final Report, Princeton University, December 21, 2006.

Nohlgren, I., 2004. Non-conventional causticization technology. Nordic Pulp Paper J. 19 (4), 470–480.

Nohlgren, I., Sinquefield, S., 2007. High temperature, high pressure, BLG with borate autocausticizing. In: International Chemical Recovery Conference, Quebec City, 2007.

Pettersson, K., 2011. Black Liquor Gasification-Based Biorefineries—Determining Factors for Economic Performance and CO_2 Emission Balances Heat and Power Technology (Ph.D. thesis). Department of Energy and Environment, Chalmers University of Technology, Göteborg, Sweden.

Sinquefield, S., Iisa, K., Fan, T., 2007. A sulfurtolerant catalyst for tar destruction. In: International Chemical Recovery Conference, Quebec City, 2007.

Sricharoenchaikul, V., 2001. Fate of Carbon Containing Compounds from Gasification of Kraft Black Liquor with Subsequent Catalytic Conditioning of Condensable Organics (Ph.D. dissertation). GA Institute of Technology, Atlanta, 2001.

Vetrov, S.V., 2009. Problems of innovation implementation in Russia. <http://conf.bstu.ru/conf/docs/0033/0756.doc>.

CHAPTER 5

Environmental Impact of Black Liquor Gasification

Biofuels which are produced from a BLG process excel in terms of well-to-wheel (WTW) carbon dioxide emission reduction and energy efficiency. Extensive European study performed by the research institutes of the auto and refinery industries and the Joint Research Centre of the European Commission confirmed this. The study was conducted with many different feedstocks, conversion processes, and fuel products. Synthetic diesel and DME produced from forest harvest residues over the BLG route both showed among the highest WTW greenhouse gas (GHG) reduction and energy efficiency. The total available black liquor volume in the United States with the conversion efficiency of this process is equivalent to approximately 5 billion gallons per year as ethanol. The renewable fuels standard calls for 16 billion gallons of cellulosic biofuels by 2022. Therefore, this route can give a significant contribution to meeting this target.

The pulp and paper companies in the United States today are meeting severe competition from low-cost producers overseas and from alternative solutions in both packaging and printed media. Mill operators and their investors now have a feasible option for breathing new life into the industry by transforming mills into biorefineries that use this fuels-from-the-forest process. This transformation completely changes a pulp mill's competitive position by adding 30−50% of profitable revenue with the typical 25−40% internal rate of return. It also makes needed reinvestment possible by replacing aged recovery boilers with high-maintenance costs and low performance. The fuels plant investment can also be used to provide additional recovery capacity allowing for higher pulp production in many cases. Mills producing only 500 tons of BLS per day are viable as fuels-from-the-forest biorefineries are using this method. Most mills are significantly larger. Such a biorefinery mill would produce upwards of 8 million gallons a year of green motor fuel calculated as gasoline equivalents at the minimum capacity size. In a mill investing in second-generation biofuels technology, jobs are not only preserved but also additional jobs are created, mainly for the extraction of biomass from the forest as well as to operate and maintain the biofuels plant. Other economic and public opinion benefits are also significant such as possible tax benefits and air emissions reductions.

According to Richard J. LeBlanc, CEO of Chemrec AB and its North American subsidiary, Chemrec USA, typical capital investment for a biorefinery project that uses fuels from the forest is $200–$400 million, depending on plant size and the costs to interconnect to the mill. While investment scenarios can vary, a common one is collaboration of funding from the technology provider, the mill itself, investors, and state and federal grants. The BLG industry is vigorously pursuing federal and state grants and loan guarantees to ramp up this technology as soon as possible to large-scale commercial capacity. As a source of ultra-clean, renewable motor fuels, the black liquor biomass gasification route that transforms pulp and paper mills into biorefineries is standing up to critical scrutiny as a viable and practical way of producing alternative, renewable energy, while making good use of the land and being gentle to the environment.

Forest biorefinery utilizing gasification in a BLGCC configuration rather than a Tomlinson boiler is predicted to produce significantly fewer pollutant emissions due to the intrinsic characteristics of the BLGCC technology. Syngas cleanup conditioning removes a considerable amount of contaminants and gas turbine combustion is more efficient and complete than boiler combustion. There could also be reductions in pollutant emissions and hazardous wastes resulting from cleaner production of chemicals and fuels that are now manufactured using fossil energy resources. In addition, it is generally accepted that production of power, fuels, chemicals, and other products from biomass resources creates a net zero generation of carbon dioxide (a GHG), as plants are renewable carbon sinks. A key component of the forest biorefinery concept is sustainable forestry. The forest biorefinery concept utilizes advanced technologies to convert sustainable woody biomass to electricity and other valuable products, and would support the sustainable management of forest lands (Farmer, 2005). In addition, the forest biorefinery offers a productive value-added use for renewable resources such as wood thinning and forestry residues and also urban waste (Miller et al., 2005; Mabee et al., 2005).

BLG whether conducted at high or low temperatures is still superior to the current recovery boiler combustion technology (Bajpai, 2008, 2012). The thermal efficiency of gasifiers is estimated to be 74% compared to 64% in modern recovery boilers, and the IGCC (integrated gasification and combined cycle) power plant could potentially generate twice the electricity output of recovery boiler power plants given the same amount of fuel (Farmer and Sinquefield, 2003; Dance, 2005). While the electrical production ratio of conventional recovery boiler power plants is 0.025−0.10 MWe/MWt, the IGCC power plant can produce an estimated 0.20−0.22 MWe/MWt (Farmer and Sinquefield, 2003; Sricharoenchaikul, 2001). This increase in electrical efficiency is significant enough to make pulp and paper mills potential exporters of renewable electric power. Alternatively, pulp mills could become manufacturers of bio-based products by becoming biorefineries. Additionally, the new technology could potentially save more than 100 trillion Btu of energy consumption annually, and within 25 years of implementation, it could save up to 360 trillion Btu per year of fossil fuel energy (Larsen et al., 2003). The new technology also offers the benefits of improved pulp yields if alternative pulping chemistries are included and reductions in solid waste discharges. Also, the process is inherently safer because the gasifier does not contain a bed of char smelt unlike in recovery boilers, which reduces the risk of deadly smelt−water explosions (Sricharoenchaikul, 2001).

IGCC power plants will reduce wastewater discharges at pulp and paper mills, even though they most likely will not significantly impact water quality (Larsen et al., 2003). Also, IGCC power plants will reduce cooling water and make-up water discharges locally at the mill, and because the efficient gasifiers will cause grid power reductions, substantial reductions in cooling water requirements at central station power plants will also occur (Larsen et al., 2003). Central station power plants have large water requirements for cooling towers in order to provide grid power to customers. Overall, the implementation of IGCC power plants will cause net savings in cooling water requirements and net reductions in wastewater discharges.

The most significant environmental impact caused by BLG will occur in air emissions. Compared to the current recovery technology, the IGCC system could cause low emissions of many pollutants, such as SO_2, NO_x, CO, Volatile organic compounds (VOCs), particulates and

Total reduced sulphur (TRS) gases, and overall reductions in CO_2 emissions. Even with improved add-on pollution control features, the recovery boiler system still causes higher overall emissions than the IGCC system (Larsen et al., 1998, 2003). Table 5.1 shows a list of different emissions and their qualitative environmental impact, along with relative emission rates for both recovery boilers and gasifiers.

Because the biomass sources at pulp and paper mills are sustainably grown, a BLG-based IGCC plant or biorefinery would transfer smaller amounts of CO_2 to the atmosphere as compared to using fossil fuels. The vast majority of CO_2 emitted would be captured from the atmosphere for photosynthesis and used for replacement biomass growth producing O_2 (Larsen et al., 2003). According to Larsen, if the pulp and paper industry converts 1.6 quads of total biomass energy to electricity, 130 billion kWh per year of electricity could be generated. This electricity generation in a BLG-based IGCC plant could displace net CO_2 emissions by 35 million tons of carbon per year within 25 years of implementation. Within 25 years of implementation, the IGCC could displace 160,000 net tons of SO_2, since most of the SO_2 produced in the process would be absorbed during H_2S recovery. Moreover, the overall reduction of TRS gases (i.e., H_2S) using gasification technology will also reduce odor, which will improve public acceptance of pulp and paper mills, particularly in populated areas. The forest biorefinery may represent a viable route to an

Table 5.1 Emission Rates for Recovery Boilers and Gasifiers			
Pollutant	**Relative Environmental Impact**	**Relative Emission Rates with Controls on Recovery Boilers**	**Relative Emission Rates with Gasification Technology**
Nitrogen oxides	High	Medium	Very low
Sulfur dioxide	High	Low	Very low
Carbon monoxide	Low	Medium	Very low
VOCs	High	Low	Very low
Particulates	High	Low–medium	Very low
Methane	Low–medium	Low	Very low
Hazardous air pollutants	Medium–high	Low	Very low
TRS	Low	Low	Very low
Wastewater	Medium–high	Low	Very low
Solids	Very low	Low	Low

Source: *Based on Larsen et al. (2003).*

improved business model for the pulp and paper industry. Indeed, mills have now entered meaningful dialogue about the possibility for change (Lynch and Chaine, 2007). Cost reductions, mergers and acquisitions, and quality-based innovation have been the strategy in recent years for many pulp and paper companies. In order to implement the forest biorefinery opportunity, it will be essential to refocus, adopt a market-driven mentality, and support ongoing product-centric research and development. An overall change of key success factors (Procter, 2006) is critical, as is modifying the company culture to support new objectives. The key success factors for implementing the forest biorefinery should be premised to a great extent on establishing a viable and sustainable business plan. The business plan definition will be supported by a product design methodology that incorporates market analysis and competitive position determination. This product selection methodology can be used to identify commercial opportunities, quantify potential economic benefits, and determine the best biorefinery product(s) for a given mill that fills market niche and shows promise for the longer term. Supply chain management will be critical in order to maintain high margins over the longer term by optimizing value chain development.

The forest biorefinery offers a business strategy that potential forestry companies are seriously considering for improving the overall financial performance of the sector (Chambost and Stuart, 2007). However, there are considerable technology and business risks related to its implementation. These risks can be mitigated to a great extent by using systematic product and process design tools for analyzing biorefinery strategies. Chambost and Stuart (2007) have described the basis for a systematic product design methodology for rapid market analysis, suitable for evaluating the economic and commercial potential of a biorefinery project, using a set of business tools that includes market and synergy identification. The preliminary application of this methodology has illustrated that the longer term competitive advantage of companies implementing the biorefinery is unlikely to be related to technology, but rather, related to the unique supply chain that companies put in place, coupled with manufacturing flexibility.

Industry leaders, investors, policy-makers, and others are now beginning to better understand the vital role to be played by biorefineries as we move from a fossil fuel-based energy economy toward a bio-based one (Connor, 2007). When properly located and operated, the potential of an integrated forest biorefinery is believed to be huge: a very attractive and synergistic business opportunity for both the co-located pulp and paper mill and for the biorefinery itself. Biorefineries are a key pathway to our biofuture, displacing fossil fuels and supplying clean, renewable and carbon neutral energy. Biorefineries fit very well at pulp and paper mills because of their inherent ability to gather and process biomass and create energy from biomass. Forest biorefinery has significant impact on the society, environment, and the industry (forest sector).

REFERENCES

Bajpai, P., 2008. Chemical Recovery in Pulp and Paper Making. PIRA International, UK, 166pp.

Bajpai, P., 2012. Biotechnology in Pulp and Paper Processing. Springer-Verlag New York Inc., New York, NY.

Chambost, V., Stuart, P.R., 2007. Selecting the most appropriate products for the forest biorefinery. Industrial Biotechnology 3 (2), 112–119.

Connor, E., 2007. The integrated forest biorefinery: the pathway to our bio-future. In: International Chemical Recovery Conference: Efficiency and Energy Management, Quebec City, QC, May 29–June 1, 2007, pp. 323–327.

Dance, M., 2005. Hydroxide Formation and Carbon Species Distributions During High-temperature Kraft Black Liquor Gasification (M.Sc. thesis). Georgia Institute of Technology, August 2005.

Farmer, M.C., 2005. The adaptable integrated biorefinery for existing pulp mills. In: Presentation at TAPPI Engineering, Pulping, and Environmental Conference, August 28–31, Philadelphia, PA.

Farmer, M., Sinquefield, S., 2003. An external benefits study of black liquor gasification. Final Report, Georgia Institute of Technology, June 15, 2003.

Larsen, E., Kreutz, T., Consonni, S., 1998. Performance and preliminary economics of black liquor combined cycles for a range of kraft pulp mill sizes. In: International Chemical Recovery Conference, Tampa, FL, June 1–4, 1998, vol. 2, pp. 675–692.

Larsen, E., Consonni, S., Katofsky, R., 2003. A cost–benefit assessment of biomass gasification power generation in the pulp and paper industry. Final Report, Princeton Environmental Institute, October 8, 2003.

Lynch, H., Chaine, B., 2007. Capital expenditures: the cost of business. Pulp Paper Canada 108 (4), 14–16.

Miller, M., Justiniano, M., McQueen, S., 2005. Energy and Environmental Profile of the U.S. Pulp and Paper Industry. Energetics Incorporated, Columbia, MD.

Mabee, W.E., Gregg, D.J., Saddler, J.N., 2005. Assessing the emerging biorefinery sector in Canada, assessing the emerging biorefinery sector in Canada. Appl. Biochem. Biotechnol. 121–124, 765–777.

Procter, A., 2006. Key success factors: a guide for prioritizing performance improvement effort. Pulp Paper Canada 107 (7/8), 59.

Sricharoenchaikul, V., 2001. Fate of Carbon-Containing Compounds from Gasification of Kraft Black Liquor with Subsequent Catalytic Conditioning of Condensable Organics (Ph.D. dissertation). Georgia Institute of Technology, 2001.

Made in the USA
Monee, IL
07 July 2026

56545376R00059